Springer Protocols Handbooks

For further volumes:
http://www.springer.com/series/8623

Springer Protocols Handbooks collects a diverse range of step-by-step laboratory methods and protocols from across the life and biomedical sciences. Each protocol is provided in the Springer Protocol format: readily-reproducible in a step-by-step fashion. Each protocol opens with an introductory overview, a list of the materials and reagents needed to complete the experiment, and is followed by a detailed procedure supported by a helpful notes section offering tips and tricks of the trade as well as troubleshooting advice. With a focus on large comprehensive protocol collections and an international authorship, *Springer Protocols Handbooks* are a valuable addition to the laboratory.

Detection of Plant Viruses

Advanced Techniques

Edited by

Aflaq Hamid and Sumiah Wani

Division of Plant Pathology, Sher-e-Kashmir University of Agriculture Sciences and Technology of Kashmir (SKUAST-K), Srinagar, Jammu and Kashmir, India

Gowhar Ali, Asif Shikari and Shabir Hussain Wani

Division of Genetics & Plant Breeding, Sher-e-Kashmir University of Agriculture Sciences and Technology of Kashmir (SKUAST-K), Srinagar, Jammu and Kashmir, India

Sahar Saleem

Division of Animal Biotechnology, Sher-e-Kashmir University of Agriculture Sciences and Technology of Kashmir (SKUAST-K), Srinagar, Jammu and Kashmir, India

Humana Press

Editors
Aflaq Hamid
Division of Plant Pathology
Sher-e-Kashmir University of Agriculture
Sciences and Technology of Kashmir (SKUAST-K)
Srinagar, Jammu and Kashmir, India

Gowhar Ali
Division of Genetics & Plant Breeding
Sher-e-Kashmir University of Agriculture
Sciences and Technology of Kashmir (SKUAST-K)
Srinagar, Jammu and Kashmir, India

Asif Shikari
Division of Genetics & Plant Breeding
Sher-e-Kashmir University of Agriculture
Sciences and Technology of Kashmir (SKUAST-K)
Srinagar, Jammu and Kashmir, India

Shabir Hussain Wani
Division of Genetics & Plant Breeding
Sher-e-Kashmir University of Agriculture
Sciences and Technology of Kashmir (SKUAST-K)
Srinagar, Jammu and Kashmir, India

Sahar Saleem
Division of Animal Biotechnology
Sher-e-Kashmir University of Agriculture
Sciences and Technology of Kashmir (SKUAST-K)
Srinagar, Jammu and Kashmir, India

Sumiah Wani
Division of Plant Pathology
Sher-e-Kashmir University of Agriculture
Sciences and Technology of Kashmir (SKUAST-K)
Srinagar, Jammu and Kashmir, India

ISSN 1949-2448 ISSN 1949-2456 (electronic)
Springer Protocols Handbooks
ISBN 978-1-0716-4389-1 ISBN 978-1-0716-4390-7 (eBook)
https://doi.org/10.1007/978-1-0716-4390-7

This Humana imprint is published by the registered company Springer Science+Business Media, LLC, part of Springer Nature.
The registered company address is: 1 New York Plaza, New York, NY 10004, U.S.A.

Foreword

Scientific advancements in the world of Plant Virology have increased manifold; however, compilation of a lot of the updated practical concepts is still limited. Plant viruses have started gaining a lot of attention due to the threat of their nature of systemic infection and complete destruction of the infected plant. Early and accurate detection and diagnosis of these viruses are crucial to manage these viruses. Plant virology laboratory in the university has done a magnificent role in pushing the academic advances in the research field of plant pathology department of the university. Books and manuals from other countries or universities are too expensive or not readily available for the students.

It gives me immense joy and pleasure to see that a plant virology manual with advances in the subject has been devised. I believe this will greatly benefit the students and scholars in the discipline. I congratulate the author of this manual, Dr. Aflaq Hamid, for this huge accomplishment and hope that he will continue to contribute to the world of scientific research and train qualified and competent scholars in the field of plant sciences.

Vice-chancellor, SKUAST-Kashmir, Srinagar, Jammu and Kashmir, India

Nazir Ahmad Ganai

Preface

Plant viruses are prevalent all around the world infecting different crops causing massive yield and quality losses. The robust nature of viruses to cause systemic infection and completely take over the plant system is what sets them apart as one of the most devastating pathogens affecting agricultural crops. The demand for higher food production is increasing daily, and to increase crop production, more efforts have to be diverted to manage the losses incurred in total agriculture production. A major part of the losses is contributed by diseases with viral diseases being one of the main parameters. Timely and efficient detection technique is key for the effective management of these viral diseases. There have been huge advancements in viral management strategies ever since the day tobacco mosaic virus was discovered. And with the continual advancements in the technology and science, new equipment with updated techniques has been developed.

This manual has been prepared by assembling together basic but crucial protocols for virus detection and diagnosis, connecting them with the recent advanced technologies developed for the detection of viral diseases. The motive behind is to create awareness and enhanced understanding of the subject and its importance with special reference to virus detection and characterisation. We believe and hope that students and scholars will benefit from this manual to ultimately improve their research studies as a whole.

Further, we would appreciate any kind of feedback and suggestions for improvement of this manual.

Operational Guidelines

There are significant and specialised physical risks surrounding every laboratory. The ultimate concern for any individual working in a laboratory should be safety. Detailed instructions must be provided to all laboratory members on how to mitigate the risks related to their work. All members need to be aware of the proper biological and chemical safety precautions while working in the laboratory.

General Guidelines

1. Always wear lab coat or apron for performing any kind of experiment in the laboratory for protection from fading chemicals or staining solutions.
2. Never eat or drink from the laboratory.
3. Always wash hands before and after entering the laboratory.
4. Ensure clean workspace by disinfecting before and after each use.
5. Never touch any kind of chemicals without the use of gloves. Avoid any kind of eye or skin contact.
6. Never wear outside shoes inside the laboratory.

7. Clothing with synthetic fibres must be avoided when working with flammable chemicals.
8. Avoid working in the laboratory alone, especially when procedures are hazardous and risky.
9. Avoid pouring chemicals down the drain or sewer for chemical waste disposal.
10. Laboratory glasses must be worn for eye protection and contact lens must be avoided especially around hazardous chemicals.
11. Never perform unauthorised work, and report any kind of accidents.
12. Always be aware of the location of first aid box, fire extinguisher and wash stations.
13. Never intentionally taste or smell chemicals.
14. Label each and every chemical solution bottle or flask that is repacked or newly formulated.

Equipment Handling

1. Use equipment only for their designated purpose.
2. Regular inspection of all equipment for wear and tear.
3. Always make sure to turn the machine off after any use.
4. Never operate new equipment without proper training or assistance.
5. Ensure proper electrical work by professionals before operation of any new equipment.

Srinagar, Jammu and Kashmir, India

Aflaq Hamid
Gowhar Ali
Asif Shikari
Shabir Hussain Wani
Sahar Saleem
Sumiah Wani

Contents

Contributors

GOWHAR ALI • *Division of Genetics and Plant Breeding, Sher-e-Kashmir University of Agriculture Sciences and Technology of Kashmir (SKUAST-K), Srinagar, Jammu and Kashmir, India*

WANI FARHANA • *Plant Virology and Molecular Pathology Laboratory, Division of Plant Pathology, Faculty of Horticulture, Sher-e-Kashmir University of Agricultural Sciences and Technology of Kashmir, Srinagar, Jammu and Kashmir, India; Division of Plant Pathology, Faculty of Horticulture, Sher-e-Kashmir University of Agricultural Sciences and Technology of Kashmir, Srinagar, Jammu and Kashmir, India*

MEHREENA FAROOQ RATHER • *Division of Plant Pathology, Faculty of Horticulture, Sher-e-Kashmir University of Agricultural Sciences and Technology of Kashmir, Srinagar, Jammu and Kashmir, India*

AFLAQ HAMID • *Division of Plant Pathology, Sher-e-Kashmir University of Agriculture Sciences and Technology of Kashmir (SKUAST-K), Srinagar, Jammu and Kashmir, India*

NULEVINO IRALU • *Plant Virology and Molecular Pathology Laboratory, Division of Plant Pathology, Faculty of Horticulture, Sher-e-Kashmir University of Agricultural Sciences and Technology of Kashmir, Srinagar, Jammu and Kashmir, India*

ANDLEEB JAVAID • *Division of Genetics and Plant Breeding, Faculty of Agriculture, Sher-e-Kashmir University of Agricultural Sciences and Technology of Kashmir, Baramulla, Jammu and Kashmir, India*

LUBNA JAVAID • *Division of Genetics and Plant Breeding, Faculty of Agriculture, Sher-e-Kashmir University of Agricultural Sciences and Technology of Kashmir, FoA Wadura, Baramulla, Jammu and Kashmir, India*

FIZALAH KAWOOSA • *Department of Animal Biotechnology, Faculty of Veterinary Sciences and Animal Husbandry, Sher-e-Kashmir University of Agricultural Sciences and Technology of Kashmir, Ganderbal, Jammu and Kashmir, India*

BIKASH MANDAL • *Advanced Centre for Plant Virology, Division of Plant Pathology, Indian Agricultural Research Institute, New Delhi, Delhi, India*

DASARI MEGHANATH • *Plant Virology and Molecular Pathology Laboratory, Division of Plant Pathology, Faculty of Horticulture, Sher-e-Kashmir University of Agricultural Sciences and Technology of Kashmir, Srinagar, Jammu and Kashmir, India*

INSHA MEHRAJ • *Plant Virology and Molecular Pathology Laboratory, Division of Plant Pathology, Faculty of Horticulture, Sher-e-Kashmir University of Agricultural Sciences and Technology of Kashmir, Srinagar, Jammu and Kashmir, India*

INSHA MEHRAJ • *Faculty of Veterinary Sciences and Animal Husbandry, Department of Animal Biotechnology, Sher-e-Kashmir University of Agricultural Sciences and Technology of Kashmir, Ganderbal, Jammu and Kashmir, India*

SAJAD UN NABI • *ICAR-Central Institute of Temperate Horticulture, Srinagar, Jammu and Kashmir, India*

HANU R. PAPPU • *Department of Plant Pathology, Washington State University, Pullman, WA, USA*

ISHRAT RAFIQ • *Division of Genetics and Plant Breeding, Faculty of Agriculture, Sher-e-Kashmir University of Agricultural Sciences and Technology of Kashmir, Baramulla, Jammu and Kashmir, India*

SHAHJAHAN RASHID • *Plant Virology and Molecular Pathology Laboratory, Division of Plant Pathology, Faculty of Horticulture, Sher-e-Kashmir University of Agricultural Sciences and Technology of Kashmir, Srinagar, Jammu and Kashmir, India*

MARIYA RATHER • *Plant Virology and Molecular Pathology Laboratory, Division of Plant Pathology, Faculty of Horticulture, Sher-e-Kashmir University of Agricultural Sciences and Technology of Kashmir, Srinagar, Jammu and Kashmir, India*

ANIRBAN ROY • *Advanced Centre for Plant Virology, Division of Plant Pathology, Indian Agricultural Research Institute, New Delhi, Delhi, India*

SAHAR SALEEM • *Division of Animal Biotechnology, Sher-e-Kashmir University of Agriculture Sciences and Technology of Kashmir (SKUAST-K), Srinagar, Jammu and Kashmir, India*

ASIF BASHIR SHIKARI • *Division of Genetics and Plant Breeding, Sher-e-Kashmir University of Agricultural Sciences and Technology of Kashmir, Srinagar, Jammu and Kashmir, India*

SHAMBHAVI SINGH • *Division of Genetics and Plant Breeding, Faculty of Agriculture, Sher-e-Kashmir University of Agricultural Sciences and Technology of Kashmir, Baramulla, Jammu and Kashmir, India*

M. SOUMYAKALA • *Plant Virology and Molecular Pathology Laboratory, Division of Plant Pathology, Faculty of Horticulture, Sher-e-Kashmir University of Agricultural Sciences and Technology of Kashmir, Srinagar, Jammu and Kashmir, India*

KOTTAM SUSHMA • *Department of Plant Pathology, College of Horticulture, Dr. YSRHU, Bhimavaram, Andhra Pradesh, India*

RAO THULASI • *Department of Fruit Science, College of Horticulture, Dr. YSRHU, Bhimavaram, Andhra Pradesh, India*

PARVAIZ ULLAH • *Plant Virology and Molecular Pathology Laboratory, Division of Plant Pathology, Faculty of Horticulture, Sher-e-Kashmir University of Agricultural Sciences and Technology of Kashmir, Srinagar, Jammu and Kashmir, India*

SUMIAH WANI • *Division of Plant Pathology, Sher-e-Kashmir University of Agriculture Sciences and Technology of Kashmir (SKUAST-K), Srinagar, Jammu and Kashmir, India*

INSHA YAQOOB • *Division of Vegetable Science, Faculty of Horticulture, Sher-e-Kashmir University of Agricultural Sciences and Technology of Kashmir, Srinagar, Jammu and Kashmir, India*

Chapter 1

Introduction

Insha Mehraj, Sumiah Wani, Nulevino Iralu, Hanu R. Pappu, and Aflaq Hamid

Abstract

The vast kingdom of plants, particularly edibles, serving huge in the nutrition of mankind are at a risk of infestation by various pathogens resulting in a number of diseases that are known today and yet many unknown left to be discovered, as we progress further. The pathogens commonly known to infect plants belong mainly to groups of fungi, bacteria, and viruses. Viruses, as we have come to know are unique in their architecture and paradoxical with respect to the level of damage they cause despite their tiny size. History stands as testament to the menace caused by viruses in the plant world especially to the crops consumed worldwide. Thus, the initial step to deal with this problem is by devising a way, a method, or a technique to confirm its presence. This aspect is largely dealt with in detection and diagnostic studies. Unlike other pathogens, detection of plant viruses is primarily based on molecular tools. Tracing back the developments in diagnostic studies, we have come far from earliest detection methods based solely on symptomatology, indicator plants, and microscopy to many robust and accurate molecular tools like enzyme-linked immunosorbent assay (ELISA), polymerase chain reaction with its advanced forms like real-time quantitative PCR (qPCR) droplet digital PCR (ddPCR), immununocapture PCR (IC-PCR), among others. The emerging technique of CRISPR-Cas is proving promising not only in diagnostics but management as well. This book aims to discuss the essential detection methods used in a plant virology laboratory while also deliberating about the newest developing portable techniques. This chapter gives a brief introductory lesson about Viruses discussing its architecture, taxonomy, symptomatology, transmission, and a quick preview of detection and diagnostics.

Key words Pathogens, Virus, Architecture, Taxonomy, Symptomatology, Transmission, Detection, and Diagnostics

1 Introduction

In the diverse world of microorganisms, every creature serves a purpose and as humans standing at the highest level of intellect, our concern is primarily focused on the usefulness of these creatures in the service of mankind and nature collectively. Microbes, though invisible to the naked eyes, are much capable, not only in terms of rendering beneficial roles but also as a menace, proving an everyday challenge to man. Narrowing this down to plants and crops

Aflaq Hamid et al. (eds.), *Detection of Plant Viruses: Advanced Techniques*, Springer Protocols Handbooks,
https://doi.org/10.1007/978-1-0716-4390-7_1,

specifically, a number of microbes jeopardize the production and quality of agricultural crops. Such organisms (not necessarily microscopic but mostly happen to be so), which cause diseases in their hosts, are referred to as pathogens. The study of such organisms is dealt with in a distinguished field of agricultural sciences called plant pathology. A number of pathogens infecting crops are known, which include fungi, bacteria, nematodes, viruses, and higher parasitic plants like *Cuscuta* sp. Zooming in, more particularly into one of these pathogens, the virus is an interesting dive. Viruses stand exceptional in this category for their distinct nature and behavior.

Viruses, in their native form are noncellular microscopic particles, which show infectious activity only in their respective living hosts and are therefore rightly stated as obligate intracellular parasites, capable of infecting eukaryotes, bacteria, and archaea as well as other viruses [1, 2]. Roger Hull, a prominent name in the field of plant virology profoundly describes virus as a set of one or more nucleic acid template molecules normally encased in a protective coat or coats of protein or lipoproteins that is able to organize its own replication only within suitable host cells. The coat surrounding the virus particle is called capsid, protecting the nucleic acid from the environment. Most plant viruses have icosahedral or helical capsid structure, while a few may have complex virion architecture [3]. The complexity of viruses is paradoxical with their tiny size, and the level of damage caused by them is undeniable, viruses are thus referred to as a box of paradoxes [4]. The ultra-minute structure of viruses has a size ranging from 20 nm to 2000 nm [5] and their genetic material is composed of strands of either RNA or DNA and can assume one of the following configurations:

1. Single-stranded RNA with positive polarity (ssRNA+).
2. Single-stranded RNA with negative polarity (ssRNA−).
3. Double-stranded RNA (dsRNA).
4. Single-stranded DNA (ssDNA).
5. Double-stranded DNA (dsDNA).

Plant viruses display abundance in their diversity and for convenience in their study; scientists have attempted to classify them into groups, families, genera, and species. This is the modern system of classifying viruses and is studied under a designated field of science called taxonomy, which encompasses the aspects of identification, classification, and nomenclature of viruses. Virus taxonomy officially began in 1966 under the activity of 43 virologists, which lead to the creation of International Committee on Taxonomy of Viruses (ICTV) [6]. ICTV lists nearly 2000 different species of viruses infecting plants [7]. However, their host range can be either restricted to a few plant families or rather be wider, infecting a number of plants across many families. Host range of a virus refers to the number of species of plants in which it can reproduce, causing infection.

The investigation of viruses deals with some crucial aspects associated with it including symptomatology and transmission mechanisms, which aid in detection and diagnosis of viral diseases for their proper management. Viral diseases display a number of symptoms varying according to the virus and host. The symptoms commonly displayed are mosaic patterns, deformed growth, chlorosis or yellowing, stunting and leaf distortions, ringspots, and vein clearing. However, some viruses may remain latent, conveying the host asymptomatic. The spread or transmission of viruses from plant to plant occurs by a number of mechanisms which include:

1. Mechanical transmission.
2. Vegetative, graft, and dodder transmission.
3. Transmission by pollen seeds.
4. Insect and mite transmission.
5. Nematode and fungal transmission.

   The transmission through insects is the most observed and the principal insect vectors belong to the order Hemiptera (aphids, leafhoppers, whiteflies, etc.).

Viruses contribute about 50% to the emerging infectious diseases of plants worldwide, threatening global food security. These infections account for global economic losses estimated at $30 billion annually [8]. The diagnostics of viral diseases, unlike other pathogens is tedious. Viruses are not easily identifiable merely by their symptoms as their symptoms can be confused with that of physiological disorders and other diseases. Also, they cannot be detected and diagnosed by the conventional methods used for other pathogens because of their unique attributes, lying on the borderline of living and nonliving organisms. However, the developing molecular techniques like enzyme-linked immunosorbent assay (ELISA) and polymerase chain reaction (PCR)-based detection methods have made the diagnosis and characterization of plant viruses easy and reliable.

The detection methods and the diagnostic field of viral diseases are expanding gradually with developments in molecular pathology and science in entirety. This book has attempted to present a step--by-step guide of laboratory protocols employed in a plant pathology laboratory, particularly dealing with viruses. From the extraction of nucleic acid, aiding in detection assay to analysis of results, this manual encompasses the procedures and protocols required for anyone aiming to work with viruses in a plant pathology laboratory.

References

1. Raoult D, Forterre P (2008) Redefining viruses: lessons from Mimivirus. Nat Rev Microbiol 6(4):315–319
2. Desnues C, Raoult D (2012) Virophages question the existence of satellites. Nat Rev Microbiol 10(3):234–234
3. Louten J (2016) Virus structure and classification. In: Essential human virology, p 19
4. Marintcheva B (2013) A box of paradoxes: the fascinating world of viruses. Bridgewater Rev 32(2):25–28
5. Sastry KS (2013) Diagnosis and detection of plant virus and viroid diseases. In: Plant virus and viroid diseases in the tropics: volume 1: introduction of plant viruses and sub-viral agents, classification, assessment of loss, transmission and diagnosis. Springer, Dordrecht, pp 233–353
6. Fauquet CM (1999) Taxonomy, classification and nomenclature of viruses. In: Encyclopedia of virology, p 1730
7. Gaur RK, Ali A, Cheng X, Mäkinen K, Agindotan B, Wang X (2021) Plant viruses, volume II: molecular plant virus epidemiology and its management. Front Microbiol 12: 756807
8. Hilaire J, Tindale S, Jones G, Pingarron-Cardenas G, Bačnik K, Ojo M, Frewer LJ (2022) Risk perception associated with an emerging agri-food risk in Europe: plant viruses in agriculture. Agric Food Secur 11(1):21

Chapter 2

Identification of Plant Viruses Based on Symptoms

Sumiah Wani, Ishrat Rafiq, Nulevino Iralu, Dasari Meghanath, and Aflaq Hamid

Abstract

Symptomatology is the first step in detecting plant viruses without which a lot of the current advancements in plant virology would not be possible. The lack of symptoms on an infected plant will create a lot of confusion during management of the disease. Hence, the first step in detecting and studying plant viruses is to identify virus-infected plants based on symptoms. Virus-like symptoms are also observed in plants due to unfavorable abiotic factors, and hence, viral disease diagnosis solely based on symptoms is very complex because of their similarities (Lievens et al., Recent research developments in microbiology, 2005). This protocol is a basic and quick method, which can help in building the basis for identification and detection of plant viruses based on symptoms.

Key words Virus, Symptoms, Management, Plants

1 Introduction

The first indication of plant viral infection can be noted through the study of symptoms. Symptomatology is crucial for diagnosing plant viral infections because visible symptoms help identify the virus and differentiate it from other plant diseases. By observing specific symptoms, researchers can pinpoint the viral pathogen and guide appropriate control measures, ultimately improving plant health management. Due to the microscopic nature of viruses, symptomology proves to be the only means of initial detection for viral infection in the field. The infection may be systemic or localized and this can also be made out through symptomology (Fig. 1). Understanding symptomatology is paramount for delving into viral disease progression, as it reveals how viruses affect plant physiology and development over time. This knowledge helps in developing strategies for disease management and can lead to breakthroughs in creating resistant plant varieties, providing an ultimate goal of

Aflaq Hamid et al. (eds.), *Detection of Plant Viruses: Advanced Techniques*, Springer Protocols Handbooks,
https://doi.org/10.1007/978-1-0716-4390-7_2,

Fig. 1 General symptoms of plant viruses in plants (Chapter 2)

enhanced agricultural resilience. However, confirmation of the definite identity of the virus infecting the concerned plant can only be carried using other tests.

2 Materials

1. Razor blade/scissors.
2. Polythene bags.
3. Hand lens.
4. Marker.
5. Tags.

3 Methods

1. Visit the suspected infected field and inspect for visible symptoms on the plants (*see* **Note 1**).
2. Walk diagonally or in between rows, either uproot the suspected infected plants or collect the leaf, fruit, flower samples (*see* **Note 2**).
3. Store the collected samples in −80 °C until processed.

4. Carefully study and compare the symptoms and describe them in tabular form depending on the severity and insect vectors involved, if any.

3.1 Observations

S.No.	Plant specimen	Characteristics	Symptoms	Vectors

4 Notes

1. Due to the highly similar symptoms of different viruses, symptomatology cannot be used for the detection of a specific virus unless they have some prominent unique symptom to tell them apart from other viruses or pathogen infection.
2. Inspection must be carried out by trained personnel for more precision in distinguishing between viral symptoms and physiological disorders.

Reference

1. Lievens B, Grauwet TJ, Cammue BP, Thomma BP (2005) Recent developments in diagnostics of plant pathogens: a review. In: Recent research developments in microbiology, part I, vol 9, p 57–79

Chapter 3

Mechanical Inoculation of Plant Viruses

Lubna Javaid, Sumiah Wani, Nulevino Iralu, Dasari Meghanath, and Aflaq Hamid

Abstract

The technique mechanical inoculation of viruses to plants plays a crucial role in understanding how plant viruses spread and how they can be controlled. It involves grinding up infected plant material and then physically applying it to the healthy plant. Mechanical inoculation is an in vitro method of virus transmission used for a variety of assays such as to produce local lesions, propagation of viruses, assay for infectivity, in host range study, and properties of the virus. Transmission of virus through this helps scientists to study the progression of the infection and develop strategies for disease management. The different steps and requirements for the mechanical inoculation are described.

Key words Mechanical inoculation, Carborundum powder, Virus inoculum, Inoculation buffer

1 Introduction

Mechanical or sap inoculation is the simplest form of virus transmission. Mechanical inoculation is vital in plant virology for efficiently transmitting viruses to study their effects, understand their life cycles, and develop resistant plant varieties. It allows researchers to explore virus–host interactions in controlled conditions, ultimately aiding in the development of effective disease management strategies [1]. Mechanical transmission may occur by methods such as rubbing of plant parts together, contact between infected and healthy plant, and action of humans either accidentally or deliberately. For the direct inoculation of the infected sap, an abrasive needs to be dusted on the surface of the healthy leaves and subsequently rubbed with the viral inoculum using a piece of cheese cloth or with fingers. Minute injuries are caused on the leaf surface allowing the infective virus or viral RNA to enter into the tissue. The virus is successfully established in the plant cell after penetration of viral genome and leads to infection in the plant. Artificial mechanical inoculation usually occurs by grinding up infected leaf

Aflaq Hamid et al. (eds.), *Detection of Plant Viruses: Advanced Techniques*, Springer Protocols Handbooks,
https://doi.org/10.1007/978-1-0716-4390-7_3,

tissue in phosphate buffer and then rubbing the extract gently on the leaves of the plant [2]. The surface of the plant is dusted with carborundum powder for artificially making wounds to facilitate the penetration of virus particles through the cuticle and epidermis of the healthy leaf [3]. It is recommended to select test plants based on the assessment of the viruses that can be expected and include at least one plant species from the same family as the sample under investigation, e.g., *Phaseolus vulgaris, Ammi majus, Chenopodium quinoa, Nicotiana benthamiana, and N. occidentalis* have been found to be suitable species for a large variety of viruses [4].

2 Materials

2.1 Buffers and Chemicals

1. Phosphate buffer, 0.02 M, pH 7.0.
2. Inoculation Buffer,0.02 M.
3. Disinfection solution.

2.2 Materials

1. Abrasive powder (Carborundum powder/Celite).
2. Mortar and pestle.
3. Petri dishes.
4. Gloves.
5. Healthy plant.
6. Viral-infected plant.

3 Methods

1. *Inoculum preparation*:
   - Take few leaf tissues from the infected plants and grind them using mortar and pestle along with buffer, i.e., 2 mL of buffer for 1 gm of tissue (*see* **Notes 2, 3** and **5**).
   - Filter the homogenate using cheese cloth, if necessary.
   - Prepare sufficient quantity of inoculum of each virus, depending upon the number of leaves or plants to be inoculated.
   - Keep inoculum over ice bucket/mini cooler.
2. *Inoculation*:
   - Take the healthy set of plants and label them before inoculation is carried out.
   - Dust abrasive powder (Carborundum/Celite) over the leaf surface (*see* **Note 6**).

- Position the leaf firmly and rub the dusted leaf surface with the virus inoculum using the cheese cloth or directly using fingers (*see* **Note 4**).
- Note that the leaves should only be moist after inoculation and there should be no dripping of the inoculum.
- Rinse off the plant surface using a wash bottle to remove superfluous inoculum.
- Maintain one set of control plants (uninoculated plants) for comparison studies (*see* **Note 1**).
- Store the plants in an insect-free environment and study the development of symptoms.
- Maintain one set of control, i.e., uninoculated plants treated with sap from healthy leaves, or with buffer only.

3.1 Observations

Test plant	Symptoms developed after inoculation	
Phaseolus vulgaris	Local	Systemic

4 Notes

1. Always include a set of control plants inoculated with buffer only to distinguish between virus symptoms and mechanical damage.
2. Sterilize all the tools and equipment before and after use to prevent cross-contamination.
3. The concentration of the virus in the inoculum can affect the success of infection. If initial attempts fail, try adjusting the concentration.
4. Wear protective clothing and gloves to avoid skin contact with the inoculum.
5. Ensure the virus source is viable, check the host plant's susceptibility and confirm the inoculum was properly prepared.
6. If leaves are too damaged, try using less carborundum or applying the inoculum more gently.

References

1. Hull R (2009) Mechanical inoculation of plant viruses. Curr Protoc Microbiol 13(1):16–16
2. Dijkstra J, de Jager C (2012) Practical plant virology: protocols and exercises. Springer Science & Business Media
3. Alcornero A (1972) Mechanical transmission of viruses from sweet potato. Phytopathology 63: 377–380
4. EPPO Bulletin (2022) PM/153(1) Mechanical inoculation of test plants 52(3):693–703

Chapter 4

Transmission of Plant Virus by Grafting

Bikash Mandal, Nulevino Iralu, Sumiah Wani, Dasari Meghanath, and Aflaq Hamid

Abstract

Graft inoculation of plant viruses is one of the most important and commonly used experimental method for the transmission of viruses (Bhat et al., Characterization of plant viruses, 2020). Transmission of phloem-borne viruses for their study and functioning is carried out using graft inoculation. Transmission study for most of the perennial woody plant viruses can only be carried out successfully through grafting technique using indicator plants (Jones, Diagnosis of plant virus diseases. CRC Press, 2019). The following protocol provides a basic and simple mechanism for the transmission of viruses.

Key words Inoculation, Graft, Indicator plants, Stock, Scion

1 Introduction

Grafting is important in plant virology because it allows for the study of virus transmission and interactions between different plant species. It helps researchers investigate how viruses spread and affect plant responses, which can lead to the development of resistant varieties and better disease management practices [1]. Grafting is a horticultural practice where tissues of different parts of a plant are united to form a whole plant. In the case of grafting for transmission of viruses, the success of grafting the plant is not necessary. The contact of cambium layers during grafting will successfully transmit the virus from the infected virus to the healthy plant part. The most convenient method to transmit a virus with known mode of transmission is graft transmission [2, 3]. Most of the viruses associated with woody plants cannot be studied using the main plant due to their very long cropping season. Grafting provides a way to investigate virus behavior in a controlled manner, facilitating the examination of how different plant tissues and varieties respond to viral infections [4 5]. It also helps in understanding the mechanisms of virus movement within plants and the

Aflaq Hamid et al. (eds.), *Detection of Plant Viruses: Advanced Techniques*, Springer Protocols Handbooks,
https://doi.org/10.1007/978-1-0716-4390-7_4,

effectiveness of potential resistance strategies. This ultimately contributes to the development of more resilient crops and improved control measures for plant diseases. The only way to study these viruses is by grafting the infected part with an indicator plant and noting the symptom development accordingly [2, 4, 6].

2 Materials

1. Knife.
2. Infected plant part.
3. Indicator healthy plant.
4. Plastic sheet or grafting tape.
5. Blotting paper/cloth.

3 Methods

3.1 Through Budding

Usually used for perennial woody plants

1. From the infected plant, cut out the bud with a fragment of wood as well and wrap it in blotting paper or cloth to keep it moist.
2. Cut out a T-section on the bark of the indicator plant using a knife based on the size of the bud cutout.
3. Loosen the bark around the sides of the T-section using the knife.
4. Place the bud on the exposed part of T section positioning it to sit on the lower portion of the T cut properly allowing both exposed sections to touch.
5. Using the plastic sheet or grafting tape, cover the T section firmly to allow the plant to grow in insect free condition.

3.2 Through Wedge Grafting

Usually used for woody or herbaceous plants.

1. Cut the basal end of the infected plant part to be grafted into V-shape using sharp knife.
2. Choose a healthy potted plant to be used as root stock.
3. Cut off the shoot from the root stock giving it a clean horizontal cut. Split the root stock by giving another vertical cut to accommodate the V-cut scion.
4. Split the vertical cut on the root stock gently and insert the scion properly aligning the exposed parts of both the scion and root stock.

5. Cover the grafted portion of the stem with grafting tape firmly and keep a plastic cover over it and remove the plastic after a week.

4 Notes

1. Care should be taken to ensure that the grafting tools used are properly disinfected to avoid transfer of virus to any other hosts.
2. Ensure the use of ethanol wash before any step to avoid any kind of contamination.

References

1. Spanò R, Ferrara M, Gallitelli D, Mascia T (2020) The role of grafting in the resistance of tomato to viruses. Plan Theory 9(8):1042
2. Jones AT (2019) Experimental transmission of viruses in diagnosis. In: Diagnosis of plant virus diseases. CRC Press, pp 49–72
3. Roistacher CN (1991) Graft-transmissible diseases of citrus. In: Handbook for detection and diagnosis, pp 151–156
4. Bhat AI, Rao GP, Bhat AI, Rao GP (2020) Transmission through grafting and budding. In: Characterization of plant viruses, pp 45–56
5. Tedesco S, Fevereiro P, Kragler F, Pina A (2022) Plant grafting and graft incompatibility: a review from the grapevine perspective. Sci Hortic 299: 111019
6. Bos L, Bos L (1970) Symptoms of virus diseases in plants, vol 87. Centre for Agricultural Publishing and Documentation, Wageningen

Chapter 5

Isolation and Identification of Virus Strains

Nulevino Iralu, Sumiah Wani, Dasari Meghanath, and Aflaq Hamid

Abstract

Viral diagnosis involves the detection of emerging, reemerging or novel viral pathogens. This can only be accomplished using methods to isolate the viral strain from an infected sample. Isolation is considered the first step in virus characterization (Bhat and Rao, Springer protocols handbooks, 2020). The following protocol illustrates a very simple method that can aid in isolating and identifying a virus.

Key words Viral strains, Identification, Isolation, Symptoms

1 Introduction

Due to the continuous coevolution of plants and pathogens, viruses also have the tendency to mutate to keep up with its ever-changing environment. New viral strains may mutate and show symptoms differing from the parent strains. Plant species susceptible to one strain may show resistance to another. This is why identification and isolation of viral strains is necessary. Isolation and identification of virus strains are crucial for managing plant diseases effectively [1]. By pinpointing the specific virus strain, researchers can accurately diagnose infections and understand their impact on plant health. This knowledge helps in developing targeted resistant varieties and allows for precise control measures to be implemented. Additionally, isolating strains aids in tracking the spread of viruses, which is essential for managing outbreaks and preventing future issues [2]. Overall, these practices are fundamental for improving crop resilience, advancing research, and maintaining agricultural productivity. Plants can be infected with a number of viruses at the same time showing similar symptoms in field conditions [3], and hence, the in vivo isolation and identification is crucial to identify and diagnose the infection. Isolation is also important to obtain pure culture of a viral strain for characteristic studies [4].

Aflaq Hamid et al. (eds.), *Detection of Plant Viruses: Advanced Techniques*, Springer Protocols Handbooks,
https://doi.org/10.1007/978-1-0716-4390-7_5,

2 Materials

1. Two sets of virus-infected plants (e.g., BCMV and CMV) with identified strains.
2. Differential hosts.
3. Inoculation essentials.

3 Method

1. Prepare inoculum for both viruses and inoculate the viruses on the different hosts as per the standard method (*see* **Note 1**).
2. Store the inoculated plants in insect-free environment (*see* **Note 2**).
3. Record the symptoms of each strain on the plants at specified time intervals and compare the symptoms.

3.1 Observations

Fro example, Bean common mosaic virus (BCMV) (*see* **Note 3**).

Strain I	Strain II

Cucumber mosaic virus (CMV).

Strain I	Strain II

4 Notes

1. Preparation of inoculum should be done with great care making sure all precautions are taken to avoid any kind of contamination.
2. The inoculated plants must be kept protected and any kind of exposure to possible viral vectors should be avoided.

3. Each recording must be studied and noted with precision ensuring that each reading is taken accurately to avoid any errors in the symptomatology analysis.

References

1. Sankaran S, Mishra A, Ehsani R, Davis C (2010) A review of advanced techniques for detecting plant diseases. Comput Electron Agric 72(1): 1–13
2. Jeger MJ (2020) The epidemiology of plant virus disease: towards a new synthesis. Plan Theory 9(12):1768
3. Naidu RA, Hughes JDA (2003) Methods for the detection of plant virus diseases. In: Plant virology in sub Saharan Africa, pp 233–253
4. Bhat AI, Rao GP (2020) Characterization of plant viruses. In: Springer protocols handbooks, p 23

Chapter 6

Determination of Dilution End Point of Viruses

Nulevino Iralu, Sumiah Wani, Dasari Meghanath, and Aflaq Hamid

Abstract

Dilution end point is an important property of viruses, which is often exploited to determine virulence of a virus. The concentration of viruses in any infected sample will differ depending on different parameters. Studying the dilution end point of a plant virus is one of the most basic steps in plant virology to study the infectivity of the concerned virus (Naidu and Hughes, Plant virology in Sub Saharan Africa, 2003). The protocol given below is a basic and simple method to easily detect the dilution end point of a virus in any plant virology laboratory.

Key words Concentration, Virus, Infection, Dilution

1 Introduction

Determining the dilution endpoint in plant virology is essential for accurately assessing the concentration of a viral suspension needed to infect plants. This process involves diluting the virus and identifying the point at which it still causes infection, which is crucial for standardizing experimental conditions and ensuring reliable results. By establishing the dilution endpoint, researchers can control the amount of virus used in experiments, making it possible to study its effects on plants more precisely [1]. This information is vital for developing effective control measures, evaluating resistance, and improving diagnostic techniques, ultimately contributing to better management of plant viral diseases. Virus concentration in infected plants can differ and depend upon varying factors. The maximum dilution of infected sap, which remains infectious, is considered to be the dilution end point (DEP) [1]. In this case, the next dilution point will make the viral sap lose its infectivity. Dilution end point of viruses may range from 10^{-1} to 10^{-7}.

Aflaq Hamid et al. (eds.), *Detection of Plant Viruses: Advanced Techniques*, Springer Protocols Handbooks, https://doi.org/10.1007/978-1-0716-4390-7_6,

2 Materials

1. Virus-infected plants.
2. Healthy set of plants.
3. Distilled water or 0.01 M Phosphate buffer pH 7.
4. Pestle and mortar.
5. Pipette.
6. Test tubes.
7. Cotton swabs/muslin cloth.
8. Abrasive powder.

3 Method

1. Take few leaf tissues from the infected plants and grind them using mortar and pestle along with buffer or distilled water (1 g/1 mL) (*see* **Note 1**).
2. Filter the homogenate using muslin cloth.
3. Dilute the homogenate tenfold dilution series using distilled water and buffer.

Undiluted sap	10^{-1}	10^{-2}	10^{-3}	10^{-4}	10^{-5}	10^{-6}	10^{-7}
	1 mL of undiluted sap +9 mL buffer	1 mL of 10^{-1} diluted sap +9 mL buffer	1 mL of 10^{-2} diluted sap +9 mL buffer	1 mL of 10^{-3} diluted sap +9 mL buffer	1 mL of 10^{-4} diluted sap +9 mL buffer	1 mL of 10^{-5} diluted sap +9 mL buffer	1 mL of 10^{-6} diluted sap +9 mL buffer

4. Dust abrasive powder and inoculate these different diluted homogenates in half-leaf replicates of test plants in Latin Square Design (*see* **Note 2**).
5. Maintain an undiluted control.
6. Rinse the plants with diluted water using wash bottle and maintain the plants in greenhouse for infectivity assay.

3.1 Observations

	Number of lesions per half leaf (days post inoculation: dpi)						
Dilutions	3	5	10	15	20	30	Average
Undiluted							
10^{-1}							
10^{-2}							
10^{-3}							
10^{-4}							
10^{-5}							
10^{-6}							
10^{-7}							

4 Notes

1. Ensure the mortar and pestle used is autoclaved beforehand.
2. Excess of abrasive powder will lead to necrosis of leaves so ensure that only a minute amount is used for inoculation.

References

1. Dijkstra J, de Jager C (2012) Practical plant virology: protocols and exercises. Springer Science & Business Media

Chapter 7

Thermal Inactivation of Plant Viruses

Sumiah Wani, Nulevino Iralu, Andleeb Javaid, and Aflaq Hamid

Abstract

Thermal inactivation point (TIP) of viruses serves as an important aspect in understanding the physico-chemical properties of different viruses and the functioning of different components in a virus particle (Hahon and Kozikowski, J Bacteriol 81:609–613, 1961). Depending on the nature and topology of the virus, the thermal activation point will differ (Dijkstra et al, Practical plant virology: protocols and exercises, 1998). This protocol offers a simple and efficient methodology by lowering the temperature by 5 °C to test the thermal inactivation point of different plant viruses.

Key words Temperature, Viruses, Management, Thermal activation point

1 Introduction

Thermal inactivation point (TIP) of viruses is defined as the minimum temperature needed for the absolute inactivation of a virus in crude sap after heat exposure of 10 min [1]. Determining the thermal inactivation point in plant virology is vital for understanding the heat sensitivity of viruses, which helps in developing effective disinfection protocols and heat treatments. By identifying the temperature at which a virus is deactivated, researchers can design methods to eliminate viruses from plant materials and equipment, preventing the spread of disease. This knowledge is crucial for ensuring the effectiveness of heat-based treatments and for enhancing biosafety practices, ultimately contributing to better management and control of plant viral infections. Determining the thermal inactivation point helps in understanding how heat affects viral stability, which is critical for developing effective heat treatments to deactivate viruses in plant materials, soil, and equipment [2]. This process ensures that heat-based disinfection methods are effective in preventing virus transmission and controlling outbreaks. By knowing the exact temperature required to inactivate a virus, researchers can design more precise and efficient treatments,

Aflaq Hamid et al. (eds.), *Detection of Plant Viruses: Advanced Techniques*, Springer Protocols Handbooks,
https://doi.org/10.1007/978-1-0716-4390-7_7,

improving the overall management of viral diseases in agriculture [3]. Additionally, this information is crucial for the development of protocols for virus-free plant production and for enhancing the safety and efficacy of plant health interventions. TIP for plant viruses usually range from 55 to 70 °C with exceptional cases like Tobacco mosaic virus (TMV), which has 95 °C [4]. For preliminary tests of unidentified viruses, a 10 °C interval should be used, but in the case of repeated tests for identified viruses, 5 °C interval works well. Thermal inactivation point of viruses can be successfully exploited for the management of seed-borne viruses.

2 Materials Required

1. Ice bucket.
2. Infected plant sample.
3. Healthy set of plants.
4. Pipette.
5. 0.1 M Phosphate buffer (pH 7).
6. Pestle and mortar (*see* **Note 1**).
7. Muslin cloth.
8. Glass tubes.

3 Method

1. Take infected plant samples and grind them in buffer using pestle and mortar (*see* **Note 2**).
2. Filter homogenate in muslin cloth.
3. Prepare 0.5 ml aliquots using pipette and glass tubes and store in ice bucket.
4. Preheat water bath and heat each sample at temperatures between 40 and 70 °C at 5 °C intervals for 10 min. Treatment should start with the highest temperature since it is easier to control the temperature as it decreases instead of increasing it (*see* **Note 1**).
5. Immediately cool them by immersing the tubes in ice bucket.
6. Inoculate the healthy plants subsequently in half leaf replicates. Rinse the plants after inoculation and maintain them for symptom analysis.
7. Note down symptom development in terms of local lesions at regular intervals.

3.1 Observations

Temperatures (°C)	Number of lesions per half leaf (days post inoculation: dpi)						
	3	5	10	15	20	30	Average
Untreated control							
40							
45							
50							
55							
60							
65							
70							

4 Notes

1. For successful conduct of this experiment, temperatures under which the samples are being heated must be precise.
2. Ensure the mortar and pestle to be used is autoclaved beforehand.

References

1. Dijkstra J, de Jager CP, Dijkstra J, de Jager CP (1998) Determination of the thermal inactivation point. In: Practical plant virology: protocols and exercises, pp 102–104
2. Bhat AI, Rao GP (2020) Characterization of plant viruses. In: Springer protocols handbooks, p 23
3. Hahon N, Kozikowski E (1961) Thermal inactivation studies with variola virus. J Bacteriol 81(4):609–613
4. Atabekov J, Nikitin N, Arkhipenko M, Chirkov S, Karpova O (2011) Thermal transition of native tobacco mosaic virus and RNA-free viral proteins into spherical nanoparticles. J Gen Virol 92(2):453–445

Chapter 8

Bioassays of Plant Viruses

M. Soumyakala, Sumiah Wani, Nulevino Iralu, and Aflaq Hamid

Abstract

Bioassays, like local lesion assays, are a reliable method for determining the concentration of infectious plant viruses. This protocol outlines the steps involved, including sample preparation, inoculation of plant leaves, and counting the resulting lesions. The number of lesions is directly correlated with the virus concentration in the inoculum. By carefully following these procedures and applying statistical analysis, researchers can accurately quantify viral titers and compare the infectivity of different virus samples.

Key words Bioassay, Half leaf method, Latin square method, Local lesion assay

1 Introduction

Bioassays have been an essential tool in plant virology since decades. Their importance in diagnostic laboratories has persisted despite technical developments because they make it possible to identify viruses and viroids that are either poorly characterized or for which there are no appropriate molecular and serological assays [1]. Estimating the infectious virus concentrations in various preparations is crucial for plant virus research because it allows researchers to examine the elements that influence infection or calculate the specific infectivity of an isolated virus, or infectivity per unit weight. Quantitative bioassays are frequently employed for this purpose. The number of local lesions on leaves implanted with the viral preparations is used in these experiments to determine infectivity. The fundamental approach for figuring out the relative infectivity was created by Holmes [2] and is known as the assay method. The concentration of the virus in the inoculum determines how many infected sites result from mechanically inoculating leaves with a virus suspension [2]. The infected locations can be counted if they show up as distinct spots or local lesions.

All virus particles are not infectious, and not all cells are accessible to the virus particles; this is likely because the virus was not

Aflaq Hamid et al. (eds.), *Detection of Plant Viruses: Advanced Techniques*, Springer Protocols Handbooks,
https://doi.org/10.1007/978-1-0716-4390-7_8,

introduced into the cells sufficiently or the outer cell walls were not wound effectively during inoculation. As a result, virus concentrations found in an infectivity assay are never absolute but rather relative. Consequently, it is only possible to establish variations in viral concentrations between preparations in a comparative manner [3].

Holmes also demonstrated the relationship between the quantity of local lesions and the viral concentration [2]. This is due to the fact that the quantity of virus given to the leaf directly correlates with the number of lesions. Since then, interest has grown in the relationship between virus concentration and the development of local lesions, as well as the quantification of the virus titer using the local lesion assay. Because of this, researchers have created a number of models that precisely calculate viral quantities based on local lesions [4–11].

Local lesion assays can be carried out on the plant's nondetachable leaves or on detached leaves in trays or petri dishes. *Nicotiana glutinosa* and *Phaseolus vulgaris* leaves are easily examined when separated; however, cowpea (*Vigna unguiculata*) and *Chenopodium* spp. leaves decay quickly and should be left on the plant [12].

2 Materials

- Pestles and mortars.
- Muslin or cheesecloth.
- Tube-equipped low-speed centrifuge.
- 1 and 5 mL (micro)pipettes.
- Test tubes (10 milliliters).
- Parafilm membrane or foil made of aluminum.
- Petri dishes with filter paper discs or plastic trays (25 × 45 × 4 cm) filled with paper towels.
- Plastic bags (for trays to be covered).
- Use tap water to clean the bottle.
- 500 mesh carborundum powder.
- Razor blade.
- Blocks made of foam plastic (about 2 × 2 × 4 cm).
- Pasteur pipettes.
- A device for counting local lesions.
- Lightbox.
- Growth chamber or conditioned chamber.

2.1 Virus and Plant Materials

1. Two plant batches, A and B, will have their virus content measured. A standard suspension of purified virus (S) will be used.
2. Based on the findings of preliminary testing, this virus is known to cause 50–100 local lesions per leaf half under laboratory conditions.
3. Plants for the experiment should have symmetrical, homogeneous leaves that are no longer than 8 cm.

2.2 Solutions

1. The phosphate buffer should be 0.01 M, pH 7.0.
2. In one deionized water (solution K), dissolve 1.362 g of KH2P04, and in one deionized water (solution Na), dissolve 1.781 g of Na2HP04.2H20. Combine 51 mL of Na with 49 mL of K.

3 Methods

3.1 Half Leaf Method

Samuel and Bald [9] have noted that testing two virus preparations on opposing half of the same leaf is the most reliable method of detecting whether they differ in activity. If there is an actual difference in susceptibility or operator bias, the solutions should be rotated between the left and right halves. The significance of the variation between virus samples can be evaluated using analysis of variance or the Student's "t" test. Using the half-leaf technique, Beale [5] discovered that when 9–16 *Nicotiana glutinosa L.* plants were inoculated, variations in tobacco mosaic virus concentration as small as 50% could be identified, with odds of 50:1 or better. Loring [13] discovered that using 40–50 half-leaves of *Phaseolus vulgaris L.* allowed for the detection of variations of 10% or more. Twenty percent was the least change that could be easily detected when *N. glutinosa* was used as the test plant. Loring noted that these results were only reliable when the viral protein concentrations utilized were close to 10 ~ gm/cc.

3.2 Procedure

3.2.1 Preparation of Virus Samples

1. Gather leaves from the diseased plant batches A and B.
2. Squeeze the slurry into centrifuge tubes by passing it through muslin or cheesecloth pieces after homogenizing the leaves in the mortars. Maintain the same handling technique for the two batches, but keep them apart.
3. To extract big cell organelles and gather the supernatants, centrifuge the sap at a low speed (5000 g for 10 min).
4. Fill four test tubes labeled Al through A4 and four more test tubes labeled B1 through B4 with four milliliters of phosphate buffer each.

5. Fill tube Al with 1 milliliter of preparation A's supernatant, wrap the tube in aluminum foil or Parafilm, and completely mix the ingredients.
6. Using a clean pipette (tip), transfer 1 mL of the suspension from tube Al to tube A2. Cover the tube and thoroughly mix.
7. Pipette the suspensions from tubes A2 and A3 into the corresponding tubes, A4 and A3.
8. Proceed with preparation B using the same process (**steps 5**, **6**, and 7).

3.2.2 Inoculation on Assay Leaves

1. Place out eight rows of four paper towels each on the bench. Mark the towels A1-A4, B1-B4, and S (standard) in accordance with the scheme. Use pencil or waterproof ink (*see* **Notes 1** and **2**).
2. Moisten the towels with tap water; do not use too much as this could interfere with the inoculation.
3. Gather leaves from the assay host (without petioles). Cut the leaves in half along the midrib and place the halves on the wet paper towels. Use eight leaves for each of the inoculum (Al through B4).
4. Dust the leaf-halves evenly and lightly with carborundum powder (excessive abrasive absorbs too much inoculum and may cause serious injury to the leaves) (*see* **Notes 3** and **4**).
5. Using a clean pipette tip for each inoculum, pipette three drops (100–150 Ill) of each inoculum onto the base of each respective leaf half.
6. Using a clean foam plastic block for each inoculum or your fingers, gently rub the drops over the leaf halves (if using your forefinger, wipe your hands well with soap after each inoculation). Avoid using too much pressure and massaging the same spot more than twice. Make an effort to use a standard stroke pattern. One benefit of using the forefinger for inoculation is that it makes it simpler to determine the appropriate amount of pressure to apply to a given kind of leaf (*see* **Note 5**).
7. Rinse the leaf halves on the paper towel under the running water right away following the inoculation. Place the towel halves on the plastic trays, then cover the trays with plastic bags. Tightly close the bags (*see* **Note 6**).
8. Grow the leaves in a conditioned chamber with constant light at 20–23 °C. Regularly check the paper towels' level of moisture (*see* **Notes** 7 and **8**).
9. Count the lesions as soon as they become noticeable (typically 3–5 days after the vaccination). Placing the leaf halves on a light box with built-in fluorescent tubes makes it easy to count

lesions since they make the lesions stand out. Prior to the lesions coalescing, they should be counted (*see* **Note 9**).

It is possible to accurately count up to 200 lesions on each leaf half, depending on the size of the lesions and the leaf itself.

3.2.3 Latin Square Method

Youden and Beale's [14] concept of using the Latin square approach can be utilized to compare multiple preparations without having to compare them to a standard. The acquired data can undergo analysis of variance, allowing for the evaluation and removal of variations resulting from individual plant variances or leaf position. The treatments in this strategy are designed so that every treatment appears on every plant at every leaf position an equal number of times. This approach cannot be selected when the number of treatments exceeds the number of leaves. As a result, randomized incomplete block design—a modified technique—is applied.

3.2.4 Procedure

1. The assay plants' tops are removed 2 days prior to inoculation in order to increase the leaves'receptivity (*see* **Notes 1** and **2**).
2. The virus preparations to be injected on both half of each leaf should be indicated by hanging labels attached to the assay leaves'petioles. Pencil or waterproof ink can be used (*see* **Note 9**).
3. Sprinkle a thin layer of homogeneous carborundum powder to the leaves (*see* **Notes 3** and **4**).
4. Using the foam plastic block or the fingertip dipped in the inoculum, rub the viral suspensions over the leaf halves. After every inoculation, wash your hands completely or use a clean foam plastic block.
5. Use tap water to rinse the plants (*see* **Note 6**).
6. Maintain ideal growing conditions for the plants by keeping them in the greenhouse (*see* **Notes** 7 and **8**).
7. After lesions appear, use a light box and counting device to count them if needed.

4 Notes

1. Use plants that are all the same size, age, color, and nutritional state.
2. Apply the inoculum to the leaves as uniformly as you can.
3. Local lesion formation has been demonstrated to benefit from the addition of materials such as phosphate-based buffers, carborundum particles, or diatomaceous earth during inoculation.

4. Adding bentonite increases the amount of local lesions because it counteracts ribonuclease.
5. During the inoculation time, stay away from bright light and any other environment that could promote wilting.
6. As soon as the leaves are inoculated, rinse them with distilled water and for a full day before inoculation, keep the plants in low light.
7. Maintain a low temperature (18–20 °C) and constant lighting for the plants. Lower temperatures cause the lesions to shrink.
8. Minimizing variations and errors brought on by different environmental conditions is crucial when doing local lesion assays. The location of the leaf on the plant, light cycles, temperature, species and age of the local lesion host, nutritional circumstances, and inoculation technique are a few examples of these.
9. Appropriately label the plants, leaf locations, and treatments. Two to three days followed by the vaccination, count the lesions.

References

1. Roenhorst JW, Botermans M, Verhoeven JTJ (2013) Quality control in bioassays used in screening for plant viruses. EPPO Bull 43: 244–249. https://doi.org/10.1111/epp.12034
2. Holmes FO (1929) Local lesions in tobacco mosaic. Bot Gaz 87:39–55
3. Dijkstra J, de Jager CP (1998) Infectivity assay on local-lesion hosts. In: Practical plant virology, pp 67–77. https://doi.org/10.1007/978-3-642-72030-7_10
4. Bald JG (1937) The use of numbers of infections for comparing the concentration of plant virus suspensions: dilution experiments with purified suspensions. Ann Appl Biol 24:33–55
5. Bald J, Iltis R, Schneider S, Gokhale D, Desjardins P (1990) Association of logistic and Poisson models of infection with some physical characteristics of a single component plant virus. J Virol Methods 27:11–28
6. Kleczkowski A (1950) Interpreting relationships between the concentrations of plant viruses and numbers of local lesions. Microbiology 4:53–69
7. Furumoto WA, Mickey R (1967) A mathematical model for the infectivity-dilution curve of tobacco mosaic virus: theoretical considerations. Virology 32:216–223
8. Furumoto WA, Mickey R (1967) A mathematical model for the infectivity-dilution curve of tobacco mosaic virus: experimental tests. Virology 32:224–233
9. Gokhale D, Bald J (1987) Relationship between plant virus concentration and infectivity: a 'growth curve'model. J Virol Methods 18:225–232
10. Hart RG, Perez-Mendez GE (1957) A simple nonparametric method for evaluating the results of half-leaf local-lesion assays. Virology 4:130–134
11. Bawden FC (1964) Plant viruses and virus diseases. Ronald, New York
12. Roberts DA (1967) Local lesion assay of plant viruses. In: Corbett MK, Sisler HD (eds) Plant virology. University of Florida, Gainesville, pp 194–210
13. Loring HS (1937) Accuracy in the measurement of the activity of Tobacco mosaic virus protein. J Biol Chem 121:637
14. Youden W, Beale HP (1934) A statistical study of the local lesion method for estimating tobacco mosaic virus. Contrib Boyce Thompson Inst 6:437

Chapter 9

Direct Enzyme-Linked Immunosorbent Assay (ELISA) for Detection of Plant Viruses

Nulevino Iralu, Sumiah Wani, Dasari Meghanath, Sahar Saleem, and Aflaq Hamid

Abstract

Detection of plant viruses using enzyme-linked immunosorbent assay (ELISA) has proved to be among the most efficient method with high specificity. There are recurring improvements being explored for ELISA to optimize its sensitivity and specificity all the more (Tabatabaei and Ahmed, Cancer cell biology: methods and protocols. Springer, New York, 2022). ELISA is a powerful tool in plant virology due to its ability to detect and quantify specific viral proteins or antigens in plant tissues (Varma and Singh, Applied plant virology. Academic, 2020). This technique is crucial for several reasons. It provides a reliable and sensitive method for diagnosing viral infections, enabling early detection and accurate identification of viruses. ELISA is also valuable for monitoring the presence and concentration of viruses in plant populations, which helps in managing disease outbreaks and implementing control measures (Rodoni, Virus Res 141:150–157, 2009). Additionally, it aids in screening plants for viral infections before they are used in breeding programs or commercial production, thus preventing the spread of viruses (Rubio et al, Front Plant Sci 11:1092, 2020). Overall, ELISA's precision and efficiency makes it an essential tool for maintaining plant health and ensuring agricultural productivity. This protocol provides an overview of the general procedure that is followed conventionally in all kinds of laboratory as well as the different types of adaptations commonly used for the purpose of plant virus detection.

Key words ELISA, Sensitivity, Antigen, Antibody, Buffer, Assay

1 Introduction

ELISA is a plate-based immuno-assay first described by Engvall and Perlmann [5], which involving reaction of antibody and antigen to detect and quantify proteins, antibodies, hormones, and peptides. It is a solid phase assay where enzyme-linked antibodies form a complex with the substrate and is detected through color development in the target analyte [6]. The most pivotal factor of this detection strategy is the antigen-antibody reaction that is distinct and highly specific [7]. This technique is rapid and possesses high sensitivity with the capability to detect virus concentration as low as

Aflaq Hamid et al. (eds.), *Detection of Plant Viruses: Advanced Techniques*, Springer Protocols Handbooks, https://doi.org/10.1007/978-1-0716-4390-7_9,

picometer to nanometer range [8]. ELISA plays a pivotal role in plant virology due to its versatility and high sensitivity in detecting and quantifying specific viral antigens [9]. This technique is crucial for accurate diagnosis, allowing for early detection of viral infections before symptoms become widespread. By enabling the precise identification of viral strains, ELISA aids in the correct diagnosis and differentiation of plant viruses, which is essential for effective disease management and prevention [10]. Additionally, ELISA is instrumental in assessing the effectiveness of control measures and treatments by quantifying virus levels and monitoring the progress of infection [9]. It also supports large-scale screening of plant material for virus-free certification, ensuring that only healthy plants are propagated and used in agriculture [11]. This not only helps in preventing the spread of diseases but also contributes to maintaining crop yield and quality. Overall, ELISA's reliability and efficiency make it an indispensable tool for plant virologists and agriculturalists aiming to safeguard plant health and improve agricultural productivity. ELISA tests are usually performed in 96 well polystyrene plates, which effectively bind antibodies and proteins. This assay is based on two different reactions—enzymatic and immunochemical. The immunochemical reaction governs the bonding between a pathogen antigen and antibody, while the enzymatic reaction leads to the modification of the material catalyzed by an enzyme, which ultimately leads to color change. Using ELISA detection method, plant viruses like *Tobacco mosaic virus* [12]; *Apple mosaic virus, Apple chlorotic leaf spot virus, Apple stem grooving virus* [13]; *Citrus tristeza virus* [14]; *Tomato spotted wilt virus* [15]; *Sorghum mosaic virus* [16]; *Cherry virus A* [17], etc. There are multiple adaptations of ELISA, but the most commonly used variants are direct, indirect, and DAS-ELISA (double antibody sandwich). All of these follow a general protocol for the reagent preparation, biotinylated antibody, and avidin-biotin-peroxidase. Direct ELISA is the simplest form of ELISA where the antigen is coated to a multi-well plate and detection is directly carried out by an antibody conjugated with an enzyme. Direct ELISA is applicable in cases where high molecule weight antigens are to be determined [18]. This method is swift and only requires fewer steps as compared to other adaptations.

1.1 Buffer Preparation

1. *Coating buffer*: Dissolve the coating buffer contents from the ELISA kit in approximately 900 mL of distilled water and adjust the pH to 9.6. Make up the solution to 1 liter and store in refrigerator. Keep it frozen in glass bottles for long-term storage (*see* **Note 1**).
2. *Wash buffer*: Dissolve the wash buffer contents from the ELISA kit in 5 liters of distilled water. Add 2.5 mL of Tween™ 20 and adjust pH to 7.4. Store in refrigerator or keep it frozen in glass bottles for long-term storage (*see* **Note 2**).

3. *Conjugate or sample buffer*: Dissolve the conjugate buffer contents from the ELISA kit in approximately 900 mL of distilled water and adjust the pH to 7.4 using sodium hydroxide. Add 0.5 mL of Tween™ 20 and make up the solution to 1 liter. Store in refrigerator but not longer than a week or keep it frozen in aliquots for long-term storage.
4. *Substrate buffer*: Dissolve 25 mL of 5× concentrate from the ELISA kit in approximately 100 mL of water. Adjust pH to 9.8 and finally make up the solution to 125 mL. Store in refrigerator or keep it frozen in glass bottles for long-term storage.
5. *Substrate solution*: From the ELISA kit, dissolve substrate 1 mg/mL in 1× substrate buffer and use immediately.

2 Materials

1. Coating buffer (0.05 M Carbonate-Bicarbonate, pH 9.6).
2. Wash solution (50 mM Tris, 0.14 M NaCl, 0.05 Tween 20, pH 8).
3. Antigen.
4. Primary antibody.
5. Blocking solution (50 mM Tris, 0.14 M NaCl, 1% BSA, pH 8).
6. Enzyme substrate.
7. Microtiter plate.
8. Stop solution (0.18 M H_2SO_4).
9. Conjugate diluent (50 mM Tris, 0.14 NaCl, 1% BSA, 0.05% Tween 20, pH 8).

3 Methods

3.1 Coating

1. In the coating buffer, dilute the antigen to the desired concentration and transfer them to the well (100ug/well) (*see* **Note 3**).
2. Incubate coated plate for 60 min and aspirate the solution from each well.
3. Wash the wells with wash solution three times (*see* **Notes 4** and **5**).

3.2 Blocking

1. Block plates with 200 uL of blocking solution in each well and incubate for 30 min.
2. Wash the wells with wash solution three times.

3.3 Conjugation

1. Using the conjugate diluent, dilute the HRP conjugate and transfer 100 uL to each well.
2. Incubate for 60 min and repeat wash step for each well six times (*see* **Note 6**).

3.4 Enzyme Substrate Reaction

1. Prepare substrate solution as per recommendation and transfer 100 uL of substrate solution to each well.
2. Incubate for 5–30 min and use stop solution to end reaction.
3. Using an ELISA reader, read the plates at appropriate wavelength depending upon the substrate (*see* **Notes 7** and **8**).

4 Notes

1. Currently, there is multiple ELISA kits available in the market, read the protocols carefully and follow them specific to the kit. Never mix the protocols for any of the chemicals from different kits.
2. Many of the reagents contain elements that are temperature sensitive, which may come out of the solution when cold; hence, it is important to ensure that all reagents are at room temperature 30 min before every assay.
3. Accuracy can be increased by having duplicates for all points in the assay. Running replicates at all points will allow the spotting of obvious outliers, which may affect your reading average.
4. It is not recommended to pour the reagent from an aliquot back to the original bottle as this can cause contamination for future samples to be processed.
5. Always ensure the use of new tips for every separate sample and chemical to prevent cross-contamination between wells.
6. To ensure that the assay worked well and to allow for proper troubleshooting if an issue arises, it is important to include all suggested background wells in in the assay performed.
7. Make sure to follow local rules and regulations while handling and discarding any kind of biological material because they are potentially harmful materials.
8. No component from different or same kits should be mixed because it can cause potential damage and hazard.

5 Observations

1. Visual observations

Sample	Strong colour +++	Moderate ++	Mild +	No colour -

2. ELISA reader: OD value for positive samples should be double or more than the value of control wells.

OD of positive samples	
OD of test samples	
OD of healthy control	
OD of buffer wells	

ELISA Sheet

1. Tissue: ____________________ Virus: ______________________.
2. Place and date of collection: _____________________________.
3. Coating antibodies:________ Dilution:________ Date: ______.
4. Extraction of antigen:___________ Date:_________________.
5. Enzyme conjugate:_________ Dilution:_______ Date: ______.
6. Substrate: _________ Concentration:________ Date: _______.
7. No. of washing: _______________ Intervals: __________ mins.
8. Incubation time: __________________ at _______________ °C.

Reaction

Positive (⁺ve), negative (⁻ve), or absorbance at _____ nm

	1	2	3	4	5	6	7	8	9	10	11	12
A												
B												
C												
D												
E												
F												
G												
H												

References

1. Tabatabaei MS, Ahmed M (2022) Enzyme-linked immunosorbent assay (ELISA). In: Cancer cell biology: methods and protocols. Springer, New York, pp 115–134
2. Varma A, Singh MK (2020) Diagnosis of plant virus diseases. In: Applied plant virology. Academic Press, pp 79–92
3. Rodoni B (2009) The role of plant biosecurity in preventing and controlling emerging plant virus disease epidemics. Virus Res 141(2): 150–157
4. Rubio L, Galipienso L, Ferriol I (2020) Detection of plant viruses and disease management: relevance of genetic diversity and evolution. Front Plant Sci 11:1092
5. Engvall E, Perlmann P (1971) Enzyme-linked immunosorbent assay (ELISA) quantitative assay of immunoglobulin G. Immunochemistry 8(9):871–874
6. Uehara-Ichiki T, Shiba T, Matsukura K, Ueno T, Hirae M, Sasaya T (2013) Detection and diagnosis of rice-infecting viruses. Front Microbiol 4:289
7. Clark BR, Engvall E (2018) Enzyme-linked immunosorbent assay (ELISA): theoretical and practical aspects. In: Enzyme immunoassay. CRC Press, pp 167–180
8. Pankaj K (2013) Virus identification and quantification. Mater Methods 3:207. https://doi.org/10.13070/mm.en.3.20
9. Boonham N, Kreuze J, Winter S, van der Vlugt R, Bergervoet J, Tomlinson J, Mumford R (2014) Methods in virus diagnostics: from ELISA to next generation sequencing. Virus Res 186:20–31
10. Alhajj M, Zubair M, Farhana A (2023) Enzyme linked immunosorbent assay. StatPearls
11. Chalam VC, Sharma VD, Sharma R, Maurya AK (2020) Modern technologies for the diagnosis and assay of plants viruses. In: Applied plant virology. Academic Press, pp 69–77
12. Van Regenmortel MHV, Burckard J (1980) Detection of a wide spectrum of tobacco mosaic virus strains by indirect enzyme-linked immunosorbent assays (ELISA). Virology 106(2):327–334
13. Turk BA (1996) Detection of apple mosaic virus, apple chlorotic leaf spot virus and apple stem grooving virus in apple trees by enzyme-linked immunosorbent assay (ELISA) using different plant tissues. Zbornik Biotehniške Fakultete Univerze v Ljubljani, Kmetijstvo 67: 151–157
14. Bar-Joseph M, Garnsey SM, Gonsalves D, Moscovitz M, Purcifull DE, Clark MF, Loebenstein G (1979) The use of enzyme-linked immunosorbent assay for detection of citrus tristeza virus. Phytopathology 69(2):190–194
15. Gao S, Wu J (2022) Detection of tomato spotted wilt virus (TSWV) infection in plants

using DAS-ELISA and dot-ELISA. In: Plant virology: methods and protocols, pp 253–261

16. Chen H, Ali N, Lv W, Shen Y, Qing Z, Lin Y, Chen B, Wen R (2020) Comparison of IC-RT-PCR, dot-ELISA and Indirect-ELISA for the detection of sorghum mosaic virus in field-grown sugarcane plants. Sugar Tech 22:122–129
17. Wani S, Shah MD, Padder BA, Nabi SU, Hamid A, Mir MM, Ara S (2024) Occurrence of cherry virus a (CVA) infecting stone fruits in Kashmir province of northwestern Himalayan region of India. Indian Phytopathol 77(1): 139–145
18. Aydin S (2015) A short history, principles, and types of ELISA, and our laboratory experience with peptide/protein analyses using ELISA. Peptides 72:4–15

Chapter 10

Direct Antigen-Coated (DAC) Enzyme-Linked Immunosorbent Assay (ELISA) for Detection of Plant Viruses

Nulevino Iralu, Sumiah Wani, Dasari Meghanath, and Aflaq Hamid

Abstract

This chapter provides a comprehensive overview of the Direct Antigen Capture Enzyme-Linked Immunosorbent Assay (DAC ELISA), a sensitive and specific technique for detecting and quantifying antigens in various biological samples. The chapter outlines the step-by-step protocol for conducting DAC ELISA, including sample preparation, coating of plates with capture antibodies, and the detection process using enzyme-labeled antibodies. Key factors such as antibody specificity, incubation times, and washing procedure influence the assay. The Direct Antigen Capture Enzyme-Linked Immunosorbent Assay (DAC ELISA) is vital for its high sensitivity and specificity in detecting target antigens, making it a cornerstone in clinical diagnostics and research. Its ability to deliver rapid results is crucial in time-sensitive situations, such as diagnosing infectious diseases. Versatile in application, DAC ELISA is utilized across various fields. The assay provides quantitative data, essential for tracking disease progression and therapeutic responses, while remaining cost-effective and straightforward to implement, making it accessible to a wide range of laboratories. Its adaptability allows for customization based on specific antigens, enhancing its utility in diverse research scenarios. Ultimately, DAC ELISA contributes significantly to pathogen detection and diagnostics by enabling the early detection of pathogens and toxins, thereby playing a critical role in monitoring and preventing diseases. This chapter serves as a practical guide for researchers and laboratory technicians seeking to implement DAC ELISA in their work.

Key words DAC-ELISA, Antigen, Antibody, Primary, Secondary

1 Introduction

Direct-antigen coated (DAC) ELISA was first developed by Lindström and Wager in 1978 [1]. Here, the detection is carried out in two stages: after the antigen is coated into the multiwall plate, an unlabeled primary antibody specific to the antigen is applied and followed by an enzyme-linked secondary antibody. The secondary antibody is usually an anti-species, polyclonal antibody [2]. The Direct Antigen Capture Enzyme-Linked Immunosorbent Assay

Aflaq Hamid et al. (eds.), *Detection of Plant Viruses: Advanced Techniques*, Springer Protocols Handbooks,
https://doi.org/10.1007/978-1-0716-4390-7_10,

(DAC ELISA) is particularly applicable in the field of plant virology for several reasons. Firstly, it allows for the sensitive and specific detection of viral antigens in various plant tissues, enabling researchers and agronomists to identify infections early and accurately [3]. This is crucial for managing plant diseases that can significantly impact crop yield and quality. Additionally, DAC ELISA can be employed to screen large numbers of plant samples efficiently, facilitating rapid diagnostics in agricultural settings [4]. Its quantitative nature allows for the assessment of viral load, which can provide insights into disease progression and help in making informed management decisions. The assay is also adaptable for different plant viruses by using specific antibodies, making it a versatile tool for studying a wide range of viral pathogens [5]. Moreover, DAC ELISA can be integrated into routine testing protocols for seed certification and plant quarantine, helping to prevent the spread of viral diseases in crops [6]. Overall, the applicability of DAC ELISA in plant virology enhances our ability to monitor and control viral infections, ultimately supporting sustainable agriculture and food security.

2 Materials

1. Enzyme-labeled secondary antibody.
2. Polyclonal antiserum.
3. 96 well plate (8 × 12).
4. Substrate.
5. Pipette.
6. Buffers.
7. Test antigen.

3 Methods

1. *Coating*: Dilute IgG 1:1000 in coating buffer and draw 200 uL diluted IgG using pipette and add that in each well (*see* **Notes 7–10**).
    - Cover the plate with parafilm/plate cover and incubate at room temperature for 1 h (*see* **Notes 1–3**).
    - Wash plates with washing buffer for 3–4 times at 5 min intervals (*see* **Note 6**).
2. *Antiserum*: Dispense 200 uL of polyclonal antiserum IgG into each well and incubate at room temperature for 1 h.
    - Wash the plate as directed above (*see* **Note 6**).

3. *Substrate*: Add 200 uL of substrate to each well.
   - Incubate at room temperature for 60 min (*see* **Note 4**).
   - End reaction by adding 50 uL of 3 M NaOH into each well and observe the color development visually or using spectrophotometer at 405 nm (ELISA reader with single filter) or 405/492 nm (ELISA reader with dual filters (*see* **Note 5**).

4 Notes

1. Use high-quality, specific antibodies and reagents to minimize cross-reactivity and ensure sensitivity. Verify the lot numbers and expiration dates before use.
2. Employ aseptic techniques throughout the procedure to avoid contamination. This includes using sterile equipment, changing pipette tips between samples, and working in a clean environment.
3. Properly handle and store samples to preserve antigen integrity. Avoid repeated freeze-thaw cycles, which can degrade samples.
4. Maintain appropriate temperatures during incubation steps and storage of reagents and samples. Sudden temperature changes can affect assay performance.
5. Adhere strictly to incubation times to ensure reproducibility. Variations can lead to inconsistent results.
6. Perform washing steps thoroughly and consistently to remove unbound antibodies. Inadequate washing can result in high background signals and false positives.
7. Include positive and negative controls in each assay to validate results to help identify any issues with the assay process.
8. Regularly calibrate equipment such as plate readers and ensure that all reagents are standardized to minimize variability.
9. Keep detailed records of all steps taken during the assay, including reagent lot numbers, sample conditions, and environmental factors, to facilitate troubleshooting if needed.
10. Ensure that personnel conducting the assay are well trained in the technique and understand the importance of each step in the process.

References

1. Lindström P, Wager O (1978) IgG autoantibody to human serum albumin studied by the ELISA-technique. Scand J Immunol 7(5): 419–425
2. Sasikala M, Chithra KR, Solomon JJ, Rajeev G (2001) Development of DAC indirect ELISA for the rapid detection of coconut root (wilt) disease. Cord 17(02):34–34

3. Kumar PL, Jones AT, Waliyar F (2004) Serological and nucleic acid based methods for the detection of plant viruses
4. Hema M, Konakalla NC (2021) Recent developments in detection and diagnosis of plant viruses. In: Recent developments in applied microbiology and biochemistry, pp 163–180
5. Yadav N, Khurana SP (2016) Plant virus detection and diagnosis: progress and challenges. In: Frontier discoveries and innovations in interdisciplinary microbiology, pp 97–132
6. Chalam VC, Sharma VD, Sharma R, Maurya AK (2020) Modern technologies for the diagnosis and assay of plants viruses. In: Applied plant virology. Academic, pp 69–77

Chapter 11

Double Antibody Sandwich Enzyme-Linked Immunosorbent Assay (ELISA) for Detection of Plant Viruses

Sumiah Wani, Nulevino Iralu, Dasari Meghanath, and Aflaq Hamid

Abstract

This chapter focuses on the Double Antibody Sandwich Enzyme-Linked Immunosorbent Assay (DAS-ELISA), a widely used technique for the detection and quantification of specific antigens, particularly in the context of plant virology and disease diagnostics. DAS-ELISA has the distinct ability to capture antigens using two distinct antibodies for enhanced specificity and sensitivity. The chapter outlines a detailed protocol highlighting the key factors that influence assay performance, such as antibody selection, incubation times, and washing protocols to optimize results. Troubleshooting strategies are also provided to address common challenges encountered during the assay. By the end of the chapter, readers will gain practical insights into the implementation and advantages of DAS-ELISA, making it a valuable tool for researchers and practitioners in the fields of plant pathology and diagnostics.

Key words DAS-ELISA, Antigen, Antibody, Well, Virus, Sample

1 Introduction

Double antibody sandwich Enzyme-Linked Immunosorbent Assay (DAS-ELISA) is an adaptation was developed by Kato and his coworkers in 1977 [1]. In DAS-ELISA, a matched antibody pair is used where each one is specific for a different, non-overlapping part of the antigen molecule. Before coating the antigen into the wells, a first antibody is coated beforehand and then the antigen is loaded. A second antibody is then added over sandwiching the antigen between the two antibody layers [2]. This adaptation is more specific and sensitive than the other adaptations and is particularly effective in identifying low levels of viral antigens in plant tissues, making early diagnosis possible [3]. Early detection is crucial for preventing the spread of viruses within crops, thereby mitigating potential yield losses. One of the key advantages of DAS ELISA is its specificity, achieved through the use of two antibodies: one that captures the antigen and another that

Aflaq Hamid et al. (eds.), *Detection of Plant Viruses: Advanced Techniques*, Springer Protocols Handbooks,
https://doi.org/10.1007/978-1-0716-4390-7_11,

facilitates detection. This dual-antibody approach enhances the assay's ability to distinguish between closely related viruses and specific strains, which is vital in accurately diagnosing viral infections [4]. Additionally, DAS ELISA can be adapted to detect a wide range of plant viruses by using tailored antibodies, making it a versatile tool in the study of various viral pathogens that affect different crops [5]. The assay also provides quantitative data on viral load, which is essential for assessing the severity of infections and understanding disease dynamics in agricultural systems [6]. This quantitative aspect aids farmers and researchers in making informed decisions regarding disease management and crop management strategies. Furthermore, DAS ELISA can be utilized in both laboratory and field settings, allowing for rapid screening of plant samples in agricultural environments. This practicality enhances the ability to manage viral outbreaks in real time [2]. In terms of agricultural practices, DAS ELISA is instrumental in screening seeds and planting materials for viral infections, contributing to effective disease management and ensuring the health of agricultural produce [7]. Moreover, this method plays a critical role in research applications, facilitating studies on virus–host interactions, epidemiology, and the development of resistant plant varieties. By integrating DAS ELISA with other molecular techniques, researchers can gain a more comprehensive understanding of viral infections and their impacts [6]. Overall, DAS ELISA is a vital method in plant virology, significantly enhancing our ability to detect and manage viral diseases. Its application supports sustainable agricultural practices and contributes to food security by helping to maintain healthy crops in the face of viral threats.

2 Materials

1. 96 well plate (8 × 12).
2. Virus-specific Immunoglobulins (IgG).
3. Test samples.
4. Pipette.
5. Buffers.
6. Enzyme conjugated IgG.
7. Parafilm.

3 Methods

3.1 Coating

1. Dilute IgG 1:1000 in coating buffer and draw 200 uL diluted IgG using pipette and add that in each well (*see* **Note 1**).

2. Cover the plate with parafilm/ plate cover and incubate at 30 °C for 4 h or at 4 °C overnight (*see* **Note 2**).
3. Wash each well three times with 200 uL of washing buffer (*see* **Notes 3–5**).

3.2 Antigen

1. Using extraction buffer, homogenize the samples and add 200 uL of this extract to each well.
2. Cover the plate and incubate at 4 °C overnight.
3. Wash each well three times with 200 uL of washing buffer.

3.3 Conjugate

1. Dilute the conjugate using conjugate buffer at 1:1000 and pipette out 200 uL to each well.
2. Cover the plate with parafilm/plate cover and incubate for 5 h at 30 °C.
3. Wash each well three times with 200 uL of washing buffer.

3.4 Substrate

1. Take substrate pNPP (para-nitrophenylphosphate) and dissolve in substrate buffer to give final concentration of 1 mg/mL.
2. Add 200 uL of substrate to each well and incubate at room temperature in the dark (20–25 °C).
3. Monitor the color development visually or using spectrophotometer at 405 nm (ELISA reader with single filter) or 405/492 nm (ELISA reader with dual filters).
4. Measure after 30 min to 2 h (*see* **Note 6**).

3.5 Washing

1. To wash the plate, completely empty it by tapping dry and ensure that there is nothing left to avoid contamination from one well to another.
2. Each well must be washed three times using 200 uL washing buffer and completely tap it dry.
3. Finally tap the plate dry removing the remaining water using clean paper towel (*see* **Notes 7** and **8**).

4 Notes

1. It is crucial to use high-quality, specific antibodies and reagents to minimize cross-reactivity and enhance assay sensitivity.
2. Maintaining a sterile environment is imperative; therefore, aseptic techniques should be employed throughout the procedure to prevent contamination, which can lead to false positives.

3. Proper sample handling and storage are also vital to preserve the integrity of viral antigens; repeated freeze-thaw cycles should be avoided.
4. During the assay, strict adherence to incubation times and temperatures is necessary to ensure reproducibility, as variations can significantly affect outcomes.
5. Thorough washing steps are critical to remove unbound antibodies, preventing high background signals that could compromise the results.
6. Including positive and negative controls in each assay is important for validating results and identifying any potential issues.
7. Additionally, maintaining detailed documentation of all procedures, including reagent lot numbers and environmental conditions, aids in troubleshooting and ensures traceability.
8. Personnel should be well trained in the technique to follow best practices consistently, thereby enhancing the reliability of DAS ELISA results in detecting plant viruses.

References

1. Kato K, Hamaguchi Y, Okawa S, Ishikawa E, Kobayashi K, Katunuma N (1977) Use of rabbit antibody IgG bound onto plain and aminoalkylsilyl glass surface for the enzyme-linked sandwich immunoassay. J Biochem 82(1):261–266
2. Chalam VC, Sharma VD, Sharma R, Maurya AK (2020) Modern technologies for the diagnosis and assay of plants viruses. In: Applied plant virology. Academic, pp 69–77
3. Song G, Wu JY, Xie Y, Liu Y, Qian YJ, Zhou XP, Wu JX (2017) Monoclonal antibody-based serological assays for detection of Potato virus S in potato plants. J Zhejiang Univ Sci B 18(12): 1075
4. Hosseini S, Vázquez-Villegas P, Rito-Palomares-M, Martinez-Chapa SO (2018) Enzyme-linked immunosorbent assay (ELISA): from A to Z. Springer, Singapore, pp 67–115
5. Boonham N, Kreuze J, Winter S, van der Vlugt R, Bergervoet J, Tomlinson J, Mumford R (2014) Methods in virus diagnostics: from ELISA to next generation sequencing. Virus Res 186:20–31
6. Baral B (2024) DAS-ELISA-based detection and RT-PCR validation for precise and timely analysis of citrus tristeza virus infection in Nepalese citrus species. Curr Res Biotechnol 7: 100221
7. Sevik MA, Kose-Tohumcu E (2011) The ELISA analysis results in tomato (Lycopersicon esculentum Mill.) seed health testing for Tobacco mosaic virus. Zemdirbyste Agric 98(3):301–306

Chapter 12

Electron Microscopy for Plant Viral Detection

Insha Yaqoob, Sumiah Wani, Nulevino Iralu, Sahar Saleem, and Aflaq Hamid

Abstract

Electron microscopy (EM) has emerged as a vital technique for the detection and characterization of plant viruses. Its unparalleled resolution allows for the direct visualization of viral particles, facilitating the identification of viruses that are otherwise difficult to detect with traditional methods. Electron microscopy, considered by some to be an old technique, is still on the forefront of both clinical viral diagnoses and viral ultrastructure and pathogenesis studies. Using the most up-to-date electron cryomicroscopy methods and ongoing technical improvements offers a broad spectrum of applications, allowing in-depth investigation of viral impact on not only the host but also the environment. Such investigations are now close to atomic resolution. In combination with bioinformatics, the transition from 2D imaging to 3D remodeling allows structural and functional analyses that extend and augment our knowledge of the astonishing diversity in virus structure and lifestyle. This study reviews the application of electron microscopy in plant virology, highlighting the different methodologies and materials used. We discuss the preparation techniques, including negative staining, thin sectioning, immunoelectron microscopy, and combinated use of TEM and SEM that enhance viral particle visibility and identification.

Key words Electron microscopy, Scanning electron microscopy, Transmission electron microscopy, STEM, Immunoelectron microscopy, Cryoelectron microscopy

1 Introduction

Electron microscopy (EM) has long been used in the discovery and description of viruses. Organisms smaller than bacteria have been known to exist since the late nineteenth century [1], but the first EM visualization of a virus came only after the electron microscope was developed. Ernst Ruska, with his mentor Max Knoll, built the first electron microscope in 1931 as the project for his Ph.D. thesis. Eight years later, Ruska and colleagues Kausche and Pfankuch were the first to visualize viruses (tobacco mosaic virus) with the EM [2], and in 1986, Ruska shared the Nobel Prize with Binnig and Rohr, developers of the scanning tunneling electron microscope.

Aflaq Hamid et al. (eds.), *Detection of Plant Viruses: Advanced Techniques*, Springer Protocols Handbooks,
https://doi.org/10.1007/978-1-0716-4390-7_12,

The high resolving power of electron microscopy (EM) permits studies at nanometer scale, providing direct images of viruses for diagnosis and research. The term "Anthropocene" indicates the enormous impact of humankind on geology and ecology [3], and this holds true also for the virosphere and its impact. EM explores and validates new concepts in virology in the Anthropocene age. This is underlined by the emergence of new taxonomic groups comprising giant viruses and "virophages" [4] and the generation of new scientific terms such as virosphere, virome, synthetic virology, or bionic viruses [5–7]. Thus, the electron microscope is essential in identifying unknown agents of emerging diseases and potential bioterrorism incidents [8]. EM is also important in monitoring the effectiveness and the standardization of probe technology for detecting biological threats [9]. Besides its use in diagnostic virology, EM has been and continues to be valuable in elucidating mechanisms of virus attachment and replication.

1.1 Types of Electron Microscopy for Viral Detection

A number of methods are available to apply electron microscopy for plant virus detection. Scanning Electron Microscopy (SEM) and Transmission Electron Microscopy (TEM) are the two most common types of electron microscopy. TEM and SEM differ in how they work and what types of images they are able to capture. Generally, a transmission EM is used, and the preparative technique most often applied is negative staining [10]. Thin section EM, however, is essential in the further characterization of a virus, e.g., the elucidation of its interaction with the host cell [10]. SEM and TEM can be combined to improve the accuracy in virus quantification. This method of viral detection is known as Scanning Transmission Electron microscopy (STEM) [11]. Larger objects, such as baculovirus occlusion bodies, are better characterized when TEM and SEM are used in tandem [12]. High resolution of the structure of virion, as well as the localization and quantification of particles and ultrastructural features within implanted occlusion body preparations, are made possible by TEM. SEM studies, on the other hand, are perfect for measurements for comparing various isolates, high throughput sample screening, and preparation quality control [12]. STEM uses the same material and methodology as that of TEM but the sections are visualized using the back scattered electron detectors as used in SEM [13].

Immuno-electron microscopy (IEM) is based on the same serological principles as ELISA, and can be used for further virus identification during routine TEM diagnosis. IEM has the advantage that it works directly with raw serum, so no further purification of immunoglobulins or conjugation steps are required [14].

2 Materials

2.1 Material for Scanning Electron Microscopy (SEM) for Plant Viral Detection

(a) **Plant Samples**

The plant tissues like leaf, stem, root and among other which show symptoms of viral infection can be used as sample for viral detection.

(b) **Targeted Low (healthy) Tissue Control**

(c) **Chemicals and Reagents**

To be able to view a biological sample in the electron microscope, it must first be stabilized or fixed preferably in a way that the ultrastructure of the cells or tissue remain as close to the living material as possible.

- Fixatives like glutaraldehyde and paraformaldehyde are used in the study.
- Osmium tetroxide is required for post fixation.
- Special isolation media, e.g., phosphate buffer, cacodylate buffer.
- Solvents, e.g., ethanol series, acetone.
- Critical point drying agents like carbon dioxide.
- Conductive coating material like the pure metals also gold and platinum.

(d) **Equipment**

- Scanning Electron Microscope (SEM).
- Lab equipment (scalpel, forceps, petri dish).
- Preserving accessories (Vacuum desiccators, Fixation vials).
- Dehydration apparatus.
- Critical Point Dryer.
- Sputter Coater or Thermal Evaporator used for coating sample.

2.2 Material for Transmission Electron Microscopy (TEM) for Plant Viral Detection

Negative staining in electron microscopy (EM) is a common technique used to visualize plant viruses due to its ability to provide detailed images of viral particles. Below are the materials required for this method:

2.2.1 Negative Staining

(a) **Plant Tissue Sample**

Infected plant tissues showing symptoms of viral infection are used for viral diagnosis.

(b) **Extraction Buffer**

Typically, a 0.1 M phosphate buffer or 0.1 M Sodium Cacodylate buffer, pH 7.4 or distilled water can be used.

(c) **Electron Microscopy Grids**

Copper grids coated with a formvar film and carbon-stabilized (*see* **Note 1**).

(d) **Staining Agents**

Uranyl acetate (2%) or phosphotungstic acid (PTA, 1–2%, pH 6.5–7.0) (*see* **Note 2**).

(e) **Equipments**

- Ultracentrifuge.
- Pipettes and tips.
- Filter paper.
- Glass slides.
- Tweezers.
- Electron microscope (TEM).
- Vortex mixer.
- Centrifuge tubes.

2.2.2 Thin Sectioning

Thin sectioning is a method that reveals structures inside viruses and cells. Samples are fixed, embedded in resin, and sliced with a diamond knife and a microtome. The sections are carefully placed onto small carbon grids and can be processed in a variety of ways. Samples can be treated with electron dense stains such as uranyl acetate or lead citrate to provide contrast, or samples might be incubated with gold-labeled antibodies. Use of gold-labeled antibodies is a powerful technique that allows the localization of specific molecules within a virus or cell [15]. Below are the materials required for this method:

(a) **Plant Samples**

Infected plant tissue like leaves, stems, roots showing symptoms of viral infection are used as samples.

(b) **Fixatives and Reagents**

Commonly used fixative in the study is glutaraldehyde (2.5 4% in a phosphate buffer, pH 7.4).

Osmium tetroxide (1–2% in a phosphate buffer, pH 7.4) is used for post-fixation.

Phosphate-buffered saline (PBS) will used for rinsing the sample.

(c) **Dehydration Agents**

Ethanol series (30, 50, 70, 90, 100%) and acetone (optional, often used interchangeably with ethanol) will be used as dehydration agents.

(d) **Embedding Medium**

Epoxy resin (Epon or Spurr's resin) is the most widely used resin for electron microscopy.

(e) **Ultramicrotome**

Ultramicrotome is needed for cutting ultrathin sections (50–100 nm). Glass or Diamond Knives are used with the ultramicrotome for precise cutting.

(f) **Grid and Staining**

Copper grids of 3 mm diameter and.

uranyl acetate and lead citrate for staining.

(g) **Electron Microscope**.

Transmission electron microscope (TEM).

2.3 Material for Immunoelectron Microscopy (IEM) for Plant Viral Detection

(a) **Samples**

The plant tissues infected by the virus like leaf, stem, root, and among others, which show symptoms of viral infection, can be used as sample for viral detection.

(b) **Fixatives**

The fixative employed in this work are glutaraldehyde (2.5% in phosphate buffer) and paraformaldehyde (4% in phosphate buffer).

(c) **Embedding Medium**

The routine method for examining biological samples is to embed the material in plastic and cut ultrathin (~60–80 nm) sections.

Resins that are commonly used as an embedding medium are LR White, Spurr's resin.

(d) **Ultramicrotome**

Diamond knife, although expensive slightly than other knives, is preferred in sectioning very thin sections that range from 60 to 80 nm.

(e) **Antibodies**

Primary antibody specific to the viral antigen and secondary antibody conjugated with an electron-dense marker (e.g., colloidal gold) is used.

(f) **Buffers**

Phosphate-buffered saline (PBS).

Gelatin (e.g., 1% BSA, normal serum).

(g) **Staining Solutions (optional)**

Uranyl acetate.

Lead citrate.

(h) **Electron Microscope.**

Transmission Electron Microscope (TEM) having high resolution capacity is used for visualization.

2.4 Material for Cryo-EM for Plant Viral Detection

(a) **Plant Virus Samples**

Plant viruses such as Tobacco mosaic virus (TMV), Potato virus X (PVX), etc.

(b) **Cryoprotectants like glycerol or sucrose, if necessary**

(c) **Cryo-EM Grids Holey carbon grids (e.g., Quantifoil) or continuous carbon grids**

(d) **Vitrification Equipment**

Vitrobot or manual plunging devices.

(e) **Cryo-EM Microscope**

A high-end electron microscope such as Titan Krios, equipped with a direct electron detector.

(f) **Computational Tools**

Software for image processing and 3D reconstruction such as RELION, CryoSPARC, or EMAN2.

3 Methods

3.1 Methodology for Scanning Electron Microscopy (SEM) for Plant Viral Detection

(a) **Sample Collection and Preparation**

Collect plant tissues from areas showing symptoms of viral infection.

Collect healthy tissues for comparison [16].

(b) **Fixation**

Fix the sample in 2–3% glutaraldehyde in phosphate or 0.1 M cacodylate buffer (pH 7.4) for 2–4 h at room temperature or overnight at 4 °C. Rinse samples three times with the buffer for 10–15 min each (*see* **Note 3**).

(c) **Post-fixation**

Post-fixation with 1–2% osmium tetroxide in the same buffer for 1–2 h at room temperature.

Rinse again with the buffer three times for 10 min each.

(d) **Dehydration**

Dehydrate samples through a graded ethanol series (30, 50, 70, 90, 100%), spending 10–15 min in each concentration.

Alternatively, acetone can be used for dehydration.

(e) **Drying**

Use a critical point dryer with CO_2 to dry the samples, replacing the ethanol or acetone with CO_2 to avoid surface tension effects for 45–60 min.

(f) **Mounting and Coating**

Mount specimens onto stubs by using conductive carbon tape, silver paint, or paste. If you use paint or paste, it must be dried overnight before going into the Sputter Coater. Coat the samples with a thin layer of conductive material (gold, platinum) using a sputter coater or thermal evaporator to prevent charging under the electron beam (*see* **Note 4**).

(g) **SEM Imaging**

(i) SEM Setup.

Put the coated samples into the SEM chamber.

Modify the SEM settings such as voltage, magnification, as well as the working distance that is ideal for examining plant developed viruses.

(ii) Observation and Imaging:

It is important that the area of interest is initially searched at low power to bring it into view.

Increase the magnification progressively to observe detailed structures of plant tissue and viral particles.

Take clear images to be referenced later on and for analyses to be made in future.

(iii) Analysis.

Compare the SEM images of infected and healthy plant tissues.

Identify characteristic features of viral infection such as inclusion bodies, changes in cell structure, and presence of viral particles.

Use software tools for image analysis and measurement of viral particles (*see* **Note 5**).

It's important to note that while SEM can be a valuable tool, it has limitations in plant viral detection because of the following reasons:

- Focuses on Surface: SEM primarily examines the sample surface, making it less effective for detecting internal virus particles.
- Limited Differentiation: SEM might not distinguish between different types of viruses based solely on morphology.

For a more comprehensive detection and identification of plant viruses, techniques like Transmission Electron Microscopy (TEM) or immunological methods (e.g., ELISA) are often used alongside SEM.

3.2 Methodology for Transmission Electron Microscopy (TEM) for Plant Viral Detection

3.2.1 Negative Staining

(a) **Sample Preparation**

Select fresh plant tissue showing symptoms of viral infection.

Homogenize approximately 0.5–1.0 g of the tissue in 5–10 ml of extraction buffer using a mortar and pestle or an automated homogenizer (*see* **Note 6**).

(b) **Clarification**

Centrifuge the homogenate at low speed (e.g., 3000–5000 g) for 10 min to remove debris.

Collect the supernatant, which contains the viral particles.

(c) **Concentration (if needed)**

For low viral load samples, concentrate the virus by ultracentrifugation.

Layer the supernatant onto a sucrose cushion (20–30% sucrose in buffer) and ultracentrifuge at 100,000 g for 1–2 h.

Resuspend the pelleted virus in a small volume of buffer (100–200 μL).

(d) **Grid Preparation**

The carbon film becomes hydrophobic with time resulting in uneven spreading of the sample and the stain. The best result will be obtained if the grid surface is made hydrophilic prior to use. This can be done using glow discharge unit (negative charge for 20–30 s at 25 mA) to glow discharge the formvar-coated grids to make them hydrophilic.

Place a small droplet (5–10 μL) of the virus suspension on the grid using a pipette.

Allow the grid to stand for 1–2 min for adsorption of viral particles.

Once the specimen has been adsorbed on to the film surface, the excess liquid is blotted off using a filter paper.

(e) **Staining**

After carefully removing excess liquid from the grid with a piece of filter paper, a droplet (5–10 μL) of the staining agent (2% uranyl acetate or 1–2% PTA) is immediately added.

Let the stain stand for 30 s to 1 min.

Blot off the excess stain with filter paper.

Air dry the grid for a few minutes (*see* **Note** 7).

(f) **Examination**

Place the stained grid into the TEM.

Adjust the microscope settings (voltage, magnification) as needed to visualize the viral particles.

Capture images of the viral particles for documentation and analysis (*see* **Note 8**).

3.2.2 Thin Sectioning

(a) **Sample Collection and Preparation**

Collect symptomatic plant tissue and cut small pieces (1–2 mm cubes) of the tissue.

(b) **Fixation**

Immerse tissue samples in glutaraldehyde (2–3%) for primary fixation for 2–4 h at room temperature or overnight at 4 °C).

Rinse samples in PBS buffer.

Post-fix with osmium tetroxide for 1–2 h at room temperature (*see* **Note 9**).

(c) **Dehydration**

Sequentially dehydrate the samples through an ethanol series (30, 50, 70, 90, and 100%) for 10–15 min at each step.

Optionally, transition to 100% acetone if acetone is used.

(d) **Infiltration and Embedding**

Gradually infiltrate samples with a mixture of resin and ethanol or acetone.

Embed in pure resin and polymerize at 60 °C for 24–48 h.

(e) **Sectioning**

Trim the embedded blocks to expose the tissue.

Ultrathin sections (50–100 nm) are cut using an ultramicrotome with a glass knife or diamond knife.

Sections are picked up from the knife and transferred to formvar or carbon coated copper grid film to support the delicate sections (*see* **Note 10**).

(f) **Staining**

Stain sections with uranyl acetate (5–10 min) followed by lead citrate (1–2 min) to enhance contrast (*see* **Note 11**).

(g) **Examination**

The stained sections are placed on copper grids and examined using a transmission electron microscope (TEM). The TEM is adjusted to appropriate accelerating voltage (usually 80–100 kV) for optimal resolution. TEM allows visualization of the ultrastructure of plant cells and the detection of viral particles.

High-resolution images are captured and analyzed to identify the presence and structure of viral particles (*see* **Note 8**).

3.3 Methodology for Immunoelectron Microscopy (IEM) for Plant Viral Detection

IEM is a powerful tool for detecting and identifying plant viruses. It combines the high resolution of electron microscopy (EM) with the specificity of antibodies.

(a) **Fixation**

Plant tissue samples are fixed to preserve ultrastructure and antigenicity in glutaraldehyde (2–3%) in a phosphate buffer (pH 7.2–7.4) and paraformaldehyde (2–4%) in a phosphate buffer (pH 7.2–7.4) (*see* **Note 9**) [17].

(b) **Dehydration and Embedding**

Dehydrate the fixed tissues through a graded ethanol series (30, 50, 70, 90, 100%).

Embed the tissue in a suitable resin (e.g., Spurr's resin or epoxy resin) [18].

(c) **Sectioning**

Cut ultrathin sections (60–90 nm) using an ultramicrotome and collect the sections on formvar-coated copper grids [19].

(d) **Blocking**

Incubate sections in a blocking solution, 1–5% Bovine Serum Albumin (BSA) or normal serum to prevent nonspecific binding (*see* **Note 12**) [20].

(e) **Primary Antibody Incubation**

Incubate sections with a primary antibody specific to the viral antigen. The antibody concentration and incubation time need to be optimized.

Wash sections thoroughly with phosphate-buffered saline (PBS) to remove excess primary antibody (*see* **Note 13**) [21].

(f) **Secondary Antibody Incubation**

Incubate sections with a secondary antibody conjugated to colloidal gold particles (usually 10–20 nm in diameter).

Wash sections thoroughly with PBS to remove excess secondary antibody [22].

(g) **Counterstaining (Optional)**

Counterstain sections with uranyl acetate and lead citrate to enhance contrast [23].

(h) **Visualization**

Examine sections under a transmission electron microscope (TEM).

Identify viral particles by the presence of electron-dense gold labels [24].

(i) **Analysis**

Compare the size, shape, and morphology of the observed virus particles with known viral structures for identification.

The presence of virus particles decorated with the specific antibodies confirms the presence of the target virus in the plant sample.

3.4 Methodology for Cryo-EM for Plant Viral Detection

Cryo-EM enables high-resolution structural analysis of plant viruses by rapidly freezing samples. Rapid freezing of virus suspensions and examination of the vitrified samples permits examination of nonchemically altered structures [25]. Specimens are flash-frozen in liquid nitrogen, transferred to the microscope in a special cold chamber, and viewed frozen in a special transmission electron microscope equipped with a cryo stage. Many digital images are made at different tilt angles and reconstructed by a computer into a three-dimensional representation [26–30]. Digital imaging also permits coloration of different components.

While other methods have become more sensitive for viral detection, cryo-EM remains valuable for identifying unknown viruses and studying viral structures.

Cryo-EM is ideally suited for exploring the 3D structure of macromolecular viral assemblies [14].

Cryo-EM remains essential for initial identification of unknown viral agents in outbreaks and for investigating viral safety of biological products [8].

Cryo-EM has a resolution high enough to discriminate between aggregated viral proteins and structured viral particles [8]. The following are the steps in Cryo -EM for plant viral detection:

(a) **Sample Preparation**

Purification: Purify the plant virus from infected plant tissue using standard virology techniques like differential centrifugation and sucrose gradient centrifugation (*see* **Note 14**).

Concentration: Concentrate the virus using ultracentrifugation or precipitation methods to achieve the optimal concentration for cryo-EM (*see* **Note 15**).

(b) **Grid Preparation**

Grid Cleaning: Clean the cryo-EM grids using plasma cleaning to remove any contaminants.

Application of Sample: Apply a small volume (3–5 μL) of the purified virus sample to the cryo-EM grid.

Blotting: Blot the grid to remove excess liquid, leaving a thin film of the sample (*see* **Note 16**).

Vitrification: Rapidly plunge the grid into liquid ethane cooled by liquid nitrogen to vitrify the sample, preventing ice crystal formation [31] (*see* **Note 17**).

(c) **Data Collection**

Microscope Setup: Load the vitrified grid into the cryo-EM microscope, ensuring the temperature is maintained at liquid nitrogen temperatures (−196 °C) (*see* **Note 18**).

Imaging: Collect images at low electron dose conditions to minimize radiation damage. Use a direct electron detector for high-resolution imaging (*see* **Note 19**).

Data Acquisition: Acquire a series of images (micrographs) at different defocus levels for contrast enhancement (*see* **Note 20**)

(d) **Image Processing**

Motion Correction: Correct for beam-induced sample movement using software like MotionCor2.

CTF Estimation: Estimate the contrast transfer function (CTF) to correct for distortions caused by the microscope.

Particle Picking: Automatically or manually select individual virus particles from the micrographs.

2D Classification: Classify the selected particles into 2D classes to improve the signal-to-noise ratio.

3D Reconstruction: Generate an initial 3D model of the virus using software like RELION or CryoSPARC [14, 31, 32] (*see* **Note 21**).

3D Refinement: Refine the 3D model iteratively to achieve higher resolution and better structural details.

Validation: Validate the final 3D model by comparing with known structures and using tools like Fourier Shell Correlation (FSC).

(e) **Analysis and Interpretation**

Model Building: Build an atomic model based on the 3D reconstruction, if the resolution permits.

Structure Analysis: Analyze the viral structure to understand the architecture, function, and interactions of the viral components.

Comparison with Other Structures: Compare the plant virus structure with other known viral structures to infer evolutionary relationships and functional similarities (*see* **Note 22**).

4 Notes

1. The use of stable, hydrophilic carbon-coated grids leads to an efficient and stable adsorption of particles and a suitable stain distribution on the grid. Hydrophilicity can be achieved by pretreatment of the grids with polycations [10, 33] or by a glow discharge [34, 35].
2. The parallel application of two negative stains with different staining features, e.g., aqueous solutions of uranyl acetate and phosphotungstic acid, might help to visualize difficult-to-assess structures more convincingly, in particular with unfamiliar specimens [10].
3. SEM Requires meticulous preparation to preserve surface structures and prevent artifacts. Thus, proper fixation is crucial to preserve viral structures.
4. Coating with conductive material is essential to avoid charging and achieve clear images.
5. SEM requires expertise in both operation and interpretation of results.
6. Ensure that the preparation process does not damage or distort the virus particles. Handle samples gently and avoid excessive blotting or drying.
7. Uranyl acetate is radioactive and toxic; handle with care and follow institutional safety protocols. Also, operators should be trained in the safe use of electron microscopes, including radiation safety protocols. Also, if your sample is in a buffer that contains phosphate or salts, wash the grid on a droplet of water before staining with uranyl acetate as uranyl will make a salt with the phosphate.

8. Experience is required to accurately interpret the images, as artifacts and background noise can sometimes complicate the analysis.
9. Proper fixation and careful handling of samples are crucial to minimize artifacts such as extraction of cellular components or disruption of viral particles.
10. Ensuring that sections are consistently thin and free of wrinkles or folds is important for clear imaging.
11. Ensure the staining solution is fresh and filtered to avoid artifacts. Choice of stain can influence the clarity and contrast of the resulting images. Uneven staining or overstaining can obscure details. Optimization of staining protocols may be necessary.
12. Adequate blocking and washing steps are necessary to minimize background noise.
13. The success of IEM heavily relies on the quality and specificity of the antibodies used.
14. Use ultra-pure water and reagents. Work in a clean environment, preferably a clean room. Wear gloves and use sterile tools to handle samples and grids.
15. Optimize sample concentration to avoid overcrowding or sparsity on the grid. Ensure even distribution of particles.
16. Use cryo-tweezers to handle grids and avoid touching them directly. Store grids in liquid nitrogen to maintain their vitrified state.
17. Ensure the temperature is consistent and the ethane/propane does not solidify. Avoid air exposure to prevent ice contamination.
18. Ensure the cryo-stage is properly cooled and free from ice contamination. Regularly check and refill the liquid nitrogen. Handle liquid nitrogen with care to avoid burns and asphyxiation hazards. Follow lab safety protocols and use appropriate personal protective equipment (PPE).
19. Regularly calibrate the electron microscope for accurate imaging. Check the alignment of the electron beam and detectors.
20. Optimize the imaging conditions. Adjust defocus settings to enhance contrast. Use appropriate magnification to balance resolution and field of view.
21. Use high-quality algorithms for alignment and 3D reconstruction. Validate the resulting 3D model against raw data to ensure accuracy.

22. Maintain detailed records of all steps in the sample preparation, data collection, and processing. Develop and follow standard operating procedures to ensure consistency.

5 TEM, SEM, IEM, and Cryo-EM for Plant Viral Detection

5.1 Transmission Electron Microscopy (TEM)

(a) **Principle**

Uses a beam of electrons transmitted through a thin sample to generate an image.

(b) **Plant Viral Detection**

Directly visualizes virus particles based on their size and morphology.

(c) **Strength**

High resolution (nanometer scale) allows for detailed analysis of viral morphology.

(d) **Weakness**

Requires specialized sample preparation (thin sections), expensive instrumentation, and expertise in interpretation [36].

5.2 Scanning Electron Microscopy (SEM)

(a) **Principle**

Uses a focused electron beam scanned across the sample surface to generate an image.

(b) **Plant Viral Detection**

May indirectly detect virus-induced cellular changes but cannot directly visualize virus particles.

(c) **Strength**

Provides high-resolution images of the sample surface, useful for observing inclusion bodies formed by some viruses.

(d) **Weakness**

Limited resolution for visualizing individual virus particles, requires specialized sample preparation (coating), and expertise in interpretation [37].

5.3 Immunoelectron Microscopy (IEM)

(a) **Principle**

Combines TEM with specific antibodies to target and visualize virus particles.

(b) **Plant Viral Detection.**

Directly identifies and locates virus particles based on specific antigen–antibody interactions.

(c) **Strength**

High specificity for targeted viruses, combines high resolution with identification capability.

(d) **Weakness**

Requires specialized expertise in immunology and electron microscopy, antibody development can be time-consuming and expensive [38].

5.4 Cryo-Electron Microscopy (Cryo-EM)

(a) **Principle**

Cryo-EM involves freezing samples in vitreous ice to preserve their natural state and imaging them using electron microscopy. There are variations like cryo-TEM and cryo-SEM.

(b) **Plant Viral Detection**

Ideal for studying the structure of viruses at near-atomic resolution. Allows for high-resolution 3D reconstruction of virus particles.

(c) **Strength**

Allows imaging of biological samples in their native state, high-resolution, minimal sample preparation artifacts.

(d) **Weakness**

Requires specialized equipment and expertise, samples must be kept at cryogenic temperatures, time-consuming and expensive [39].

6 Choosing the Right Technique

- TEM: Ideal for detailed morphological analysis of purified virus particles or visualizing virus particles within thin sections of infected plant tissue.
- SEM: Useful for observing virus-induced changes in plant cell morphology, particularly for viruses that form inclusion bodies.
- IEM: Preferred method for specific identification and localization of targeted viruses within plant tissues.
- Cryo-EM provides near-native state imaging of biological samples with very high resolution but is expensive and complex.

7 Conclusion

- TEM offers the highest resolution but lacks specificity.
- SEM provides surface information but is limited in directly detecting viruses.
- IEM combines the strengths of TEM and antibody-based identification.
- Cryo-EM can achieve atomic-level resolution, providing detailed images of viral particles.

The most suitable technique depends on the specific needs of the investigation. For routine screening, IEM is often preferred due to its high specificity for targeted viruses. However, TEM and SEM can play valuable roles in complementary analyses.

References

1. Carter KC (2003) The rise of causal concepts of disease: case histories. Ashgate, Aldershot
2. Kausche G, Pfankuch E, Ruska H (1939) Die Sichtbarmachung von pflanzlichem Virus im Übermikroskop. Naturwissenschaften 27: 292–299
3. Crutzen PJ (2002) Geology of mankind. Nature 415:23. https://doi.org/10.1038/415023a
4. Koonin EV, Dolja VV, Krupovic M (2015) Origins and evolution of viruses of eukaryotes: the ultimate modularity. Virology 479–480:2–25. https://doi.org/10.1016/j.virol.2015.02.039
5. Suttle CA (2007) Marine viruses – major players in the global ecosystem. Nat Rev Microbiol 5:801–812
6. Guenther CM, Kuypers BE, Lam MT, Robinson TM, Zhao J, Suh J (2014) Synthetic virology: engineering viruses for gene delivery. Wiley Interdiscip Rev Nanomed Nanobiotechnol 6:548–558. https://doi.org/10.1002/wnan.1287
7. Koch C, Eber FJ, Azucena C, Forste A, Walheim S, Schimmel T et al (2016) Novel roles for well-known players: from tobacco mosaic virus pests to enzymatically active assemblies. Beilstein J Nanotechnol 7:613–629
8. Roingeard P (2008) Viral detection by electron microscopy: past, present and future. Biol Cell 100:491–501
9. Petrenko VA, Sorokulova IB (2004) Detection of biological threats. A challenge for directed molecular evolution. J Microbiol Methods 58: 147–168
10. Biel SS, Gelderblom HR (1999) Diagnostic electron microscopy is still a timely and rewarding method. J Clin Virol 13(1–2):105–119
11. Hartel P, Rose H, Dinges C (1996) Conditions and reasons for incoherent imaging in STEM. Ultramicroscopy 63:93–114
12. Gencer D, Bayramoglu Z, Nalcacioglu R, Kleespies RG, Demirbag Z, Demir I (2018) Characterisation of three Alphabaculovirus isolates from the gypsy moth, Lymantria dispar dispar (Lepidoptera: Erebidae). Turkey Biocontr Sci Technol 28:107–121
13. Blancett CD, Fetterer DP, Koistinen KA, Morazzani EM, Monninger MK, Piper AE et al (2017) Accurate virus quantitation using a scanning transmission electron microscopy (STEM) detector in a scanning electron microscope. J Virol Methods 248:136–144
14. Richert-Pöggeler KR, Franzke K, Hipp K, Kleespies RG (2019) Electron microscopy methods for virus diagnosis and high resolution analysis of viruses. Front Microbiol 9:3255
15. Debarbieux L, Fischer M, Quax, TEF (2017) Viruses of microbes. Viruses 9(9):37–52. https://doi.org/10.3390/v9090263
16. Golding SM, Kuzminov FI (2016) Preparation and use of plant samples for scanning electron microscopy. J Plant Physiol 213:18–25
17. Hayat MA (2000) Principles and techniques of electron microscopy: biological applications. Cambridge University Press
18. Bozzola JJ, Russell LD (1999) Electron microscopy: principles and techniques for biologists. Jones & Bartlett Publishers
19. Reynolds ES (1963) The use of lead citrate at high pH as an electron-opaque stain in electron microscopy. J Cell Biol 17(1):208–212
20. Slot JW, Geuze HJ (1985) A new method of preparing gold probes for multiple-labeling cytochemistry. Eur J Cell Biol 38(1):87–93
21. Roth J, Taatjes DJ (1998) Antibody techniques for immunoelectron microscopy. J Histochem Cytochem 46(6):877–880
22. Bendayan M (2001) Colloidal gold post-embedding immunocytochemistry. Prog Histochem Cytochem 36(1):37 66
23. Watson ML (1958) Staining of tissue sections for electron microscopy with heavy metals. J Biophys Biochem Cytol 4(4):475–478
24. Milne RG (1989) Immunosorbent electron microscopy in plant virus studies. Methods Virol 7:265–281
25. Adrian M, Dubochet J, Lepault J, McDowall AW (1984) Cryo-electron microscopy of viruses. Nature 308:32–36
26. Liu J, Bartesaghi A, Borgnia MJ, Sapiro G, Subramaniam S (2008) Molecular architecture of native HIV-1 gp120 trimers. Nature 455: 109–113

27. Provencher SW, Vogel RH (1988) Three-dimensional reconstruction from electron micrographs of disordered specimens. I. Method. Ultramicroscopy 25: 209–221
28. Sougrat R, Bartesaghi A, Lifson JD, Bennett AE, Bess JW, Zabransky DJ, Subramaniam S (2007) Electron tomography of the contact between T cells and SIV/HIV-1: implications for viral entry. PLoS Pathog 3:e63
29. Subramaniam S, Bartesaghi A, Liu J, Bennett AE, Sougrat R (2007) Electron tomography of viruses. Curr Opin Struct Biol 17:596–602
30. Vogel RH, Provencher SW (1988) Three-dimensional reconstruction from electron micrographs of disordered specimens II. Implementation and results. Ultramicroscopy 25:223–239
31. Goldsmith CS, Miller SE (2009) Modern uses of electron microscopy for detection of viruses. Clin Microbiol Rev 22(4):552–563
32. Richert-Pöggeler KR, Franzke K, Hipp K, Kleespies RG (2019) Electron microscopy-methods for virus diagnosis and high resolution analysis of viruses. Front Microbiol 9:3255
33. MacRae J, Srivastava M (1998) Detection of viruses by electron microscopy: an efficient approach. J Virol Methods 72(1):105–108
34. Aebi U, Pollard TD (1987) A glow discharge unit to render electron microscope grids and other surfaces hydrophilic. J Electron Microsc Tech 7(1987):29–33
35. Harris P, Cable J, Tinsley R (1997) Detection of sperm within Gyrodactylus (Platyhelminthes, Monogenea) tissues using fluorescence and transmission electron microscopy. Parasitol Res 83:533–538
36. Xie L, Hong J (2022) Transmission electron microscopic methods for plant virology. In: Wang A, Li Y (eds) Plant virology. Methods in molecular biology, vol 2400. Humana, New York
37. Haddad G, Bellali S, Fontanini A, Francis R, La Scola B, Levasseur A, Bou Khalil J, Raoult D (2020) Rapid scanning electron microscopy detection and sequencing of severe acute respiratory syndrome coronavirus 2 and other respiratory viruses. Front Microbiol 11:596180
38. Bodakuntla S, Kuhn CC, Biertümpfel C Mizuno N (2023) Cryo-electron microscopy in the fight against COVID-19—mechanism of virus entry. Frontiers in Molecular Biosciences 10:1252529
39. Cheng Y (2018) Single-particle Cryo-EM – how did it get here and where will it go. Science 361(6405):876–878

Chapter 13

Extraction of High-Quality Genomic DNA from Plants Using Modified CTAB-Based Method

Andleeb Javaid, Sumiah Wani, Nulevino Iralu, Lubna Javaid, Dasari Meghanath, Gowhar Ali, and Aflaq Hamid

Abstract

The DNA extraction from plant viruses begins with the collection of plant samples that exhibit viral infection symptoms. The tissue is then crushed into fine powder using liquid nitrogen. The next step is to lyse cell membranes and release the viral DNA using the extraction buffer, composed of cetyltrimethylammonium bromide (CTAB) and other reagents. CTAB protocol is the precise and efficient technique for DNA extraction from plant viruses for understanding plant–virus interactions and developing effective diagnostic tools. Subsequent steps involve chloroform:isoamyl-alcohol extractions to remove proteins, polysaccharides, and other contaminants. The DNA is precipitated with isopropanol, washed with ethanol, and finally resuspended in TE buffer or distilled water. The resulting DNA can be quantified using a spectrophotometer or fluorometer, measuring the yield and purity of the extracted material. This optimized CTAB-based protocol offers several advantages, including simplicity, cost-effectiveness, and reproducibility. It is adaptable to a wide range of plant species, making it a valuable tool for researchers and diagnosticians studying plant virology.

Key words DNA, CTAB buffer, Pellet, Centrifuge

1 Introduction

DNA extraction is crucial in plant virology for several reasons. It allows researchers to obtain pure and intact DNA from plant tissues, which is essential for detecting and identifying viral DNA sequences through various molecular techniques. This process facilitates the study of viral genomes, enabling detailed analysis of viral structure, replication, and variation. Accurate DNA extraction also supports the development of molecular diagnostics, such as PCR-based assays, which are vital for early and precise detection of plant viruses. Furthermore, DNA extraction is fundamental for genetic studies, including the creation of virus-resistant plant varieties and understanding plant–virus interactions [1]. Overall,

Aflaq Hamid et al. (eds.), *Detection of Plant Viruses: Advanced Techniques*, Springer Protocols Handbooks,
https://doi.org/10.1007/978-1-0716-4390-7_13,

effective DNA extraction is a cornerstone of plant virology research and contributes significantly to improving plant health and agricultural productivity. A crucial prerequisite for molecular biology applications such as polymeric chain reaction (PCR), rapid amplified polymorphic DNA (RAPD), restriction fragment length polymorphism (RFLP), and PCR-RFLP is the isolation of high quality, semi-intact genomic DNA [2]. Extracted pure DNA gives good polymerase chain reaction (PCR) products compared to DNA with lower purity, requiring several dilutions before giving a good PCR product [3]. CTAB (cetyltrimethylammonium bromide) buffer has been used in many protocols with some variations and is widely used for DNA extraction from diverse plant species and taxa. However, it includes using organic compounds and is potentially more toxic than other methods. This method is advisable for plants containing high polysaccharides and polyphenol content. The modified CTAB method is faster than the original method because it does not require the CsCl gradient and selective precipitation steps. It also utilizes less hazardous and expensive reagents and requires inexpensive laboratory equipment. Moreover, it is more readily suited to high-throughput DNA extraction [4].

Suboptimal DNA extractions are caused by the same chemical and anatomical properties that allow plants to thrive in their environment. For example, rigid cell walls support structure but trap DNA inside the cell, and their polysaccharides can obstruct downstream molecular techniques like ligation, PCR amplification, and enzymatic reactions [5]. Furthermore, secondary metabolites provide plant defense mechanisms [6]. Various secondary metabolites are present within plant cells, such as phenolic compounds, polysaccharides, alkaloids, anthocyanins, saponins, resins, and essential oils but it is very difficult to eliminate phenolic compounds and polysaccharides during the extraction process as they coprecipitate with DNA [7]. Two significant problems arise during DNA extraction because of DNA coprecipitation and polysaccharides, firstly when alcohol is added, the reaction becomes viscous, which results in inaccurate pipetted volumes and secondly, polysaccharides inhibit the downstream molecular reactions. The CTAB protocol has been modified in many ways by the addition of chemical additives to the suspension or lysis buffer to reduce the phenolic compound and polysaccharide coprecipitation [8].

1.1 Principle

- *CTAB*: Cetyltrimethyl-ammonium bromide or CTAB is a cationic detergent that constitutes a long hydrophobic hydrocarbon chain (water-repelling) and a hydrophilic head (soluble in water). Cell membranes are composed of diverse phospholipids and proteins, and CTAB can be utilized to break the cell membranes (Fig. 1) (*see* **Note 2**).

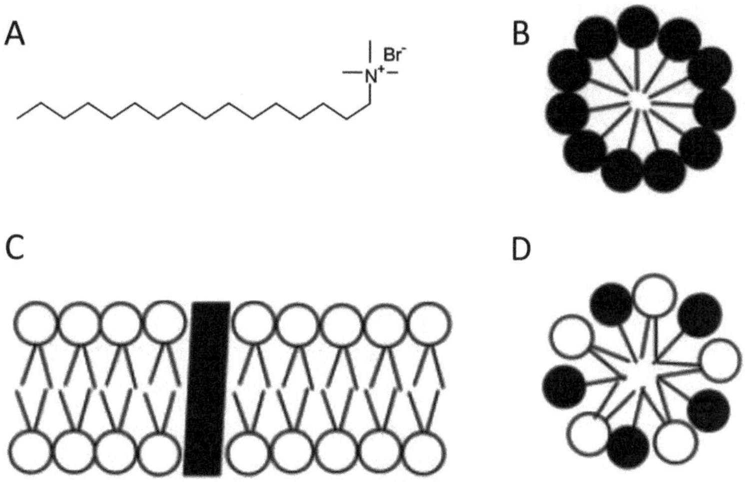

Fig. 1 (**a**). Amphipathic nature of CTAB; (**b**). CTAB forms the micelle into aqueous solution; (**c**). Biological membrane made up of amphipathic lipids with integral protein; (**d**). Formations of hybrid micelle

- *NaCl*: Salt is used to remove proteins that are bound to DNA and keep the proteins dissolved in the aqueous layer, avoiding DNA coprecipitate with proteins.
- *Tris*: The (hydroxymethyl) aminomethane is a buffer favoring the stability of DNA once it is released. The cell pH gets altered when cytoplasmic material is released, consequently, the pH change can hamper the DNA.
- *EDTA*: Ethylene diamine tetra acetic acid (EDTA) chelates divalent cations, such as Mg^{2+} and Ca^{2+}, which are cofactors of DNases and make DNases nonfunctional.
- *β-Mercaptoethanol* (BME): BME is a potent reducing agent to clean tannins and polyphenols. It favors the dissolution of globular proteins in water.
- *PVP*: Polyvinylpyrrolidone (PVP) is added to remove phenolic compounds from plant DNA extracts.
- *Phenol*: Phenol is an organic solvent, immiscible with water. It is used with chloroform and isoamyl alcohol to purify the DNA and remove proteins and polysaccharide contaminants.

2 Materials

1. Plant sample.
2. 1× CTAB buffer.
3. 1.4 M NaCl.

4. 0.8 M Tris–HCl pH 8.0.
5. 0.5 M EDTA pH 8.0 (autoclaved).
6. 0.3% 2-β-Mercaptoethanol.
7. Chloroform:isoamyl alcohol (24:1 v/v).
8. 6 M NaCl.
9. 3 M potassium acetate.
10. 100% isopropyl alcohol.
11. 70% ethanol.
12. 1× TE buffer (10 mM Tris–HCl, pH 8.0; 1 mM EDTA, pH 8.0, autoclaved).
13. Agarose (molecular grade).

3 Method

Extraction procedures for plant DNA, in general, must accomplish the following five steps:

1. *Cell wall breakdown*: Grind the tissue in dry ice or liquid nitrogen with a mortar and pestle to break down the cell walls and release the cellular contents (*see* **Note 1**).
2. *Cell membrane breakdown*: After the cell wall is broken, the cell membranes must be disrupted so that the DNA is released into the extraction buffer. Researchers accomplish this by using a detergent, usually CTAB or SDS buffers.
3. *DNA protection*: Once DNA is released into the extraction buffer, endogenous nucleases can degrade the DNA. There are some reagents used to prevent DNA degradation, such as EDTA (ethylenediaminetetraacetic acid).
4. *DNA isolation*: The DNA in the extraction buffer is isolated from other components (like proteins, polysaccharides, and phenolics). The buffer/tissue mixture is emulsified with either chloroform or phenol to denature and separate the contaminants from DNA to achieve isolation.
5. *DNA precipitation*: After removing contaminants, the DNA is precipitated with ethanol (generally 70%) and dissolved later in the water. Finally, DNA is stored at −20 °C.

3.1 Modified Protocol

1. Preheat the 1× extraction buffer in water bath at 65 °C. Add 0.3% 2-β-mercaptoethanol to the 1× CTAB extraction buffer immediately before use (*see* **Note 3**).
2. Grind 50 mg of plant samples into powder in liquid nitrogen using prechilled mortar and pestle. While still in the mortar, add 800 μL of the preheated 1× CTAB extraction buffer to the

ground plant samples and swirl gently to mix using the pestle (*see* **Note 5**).

3. Transfer the sample mixture to a 2 mL microcentrifuge tube, incubate in water bath at 60–65 °C for 1 h, mix gently every 20 min by inverting the tubes, remove the tubes from water bath and cool down to the room temperature (*see* **Notes 4**, **8** and **9**).
4. Add an equal volume of chloroform:isoamyl alcohol (24:1 v/v) and mix by slight inversion (*see* **Note 6**).
5. Centrifuge at 13,000 rpm for 15 min at room temperature.
6. Using pipette, draw the upper aqueous phase, which contains the DNA, to a new eppendorf tube.
7. Repeat the extraction **steps** (**4–6**), when necessary, until the upper aqueous phase is clear.
8. Estimate the volume of the aqueous phase (approximately 700 μL) then add half this volume (350 μL) of 6 M NaCl and mix well.
9. Add 1/10 the volume (70 μL) 3 M potassium acetate and simultaneously mix with 500 μL ice cold 100% isopropyl alcohol (approximately 2/3 the volume of the aqueous phase). Invert gently to precipitate DNA until the formation of DNA threads (*see* **Notes 10** and **14**).
10. Incubate at −20 °C for 30 min.
11. Centrifuge at 13,000 rpm for 5 min and discard the supernatant.
12. Invert the tube containing the DNA pellet on tissue paper to drain off the supernatant completely.
13. Wash DNA pellet with 500 μL of 70% ethanol and invert once (to dissolve residual salts and to increase purity of the DNA) (*see* **Notes** 7 and **12**).
14. Centrifuge at 13,000 rpm for 5 min.
15. Repeat the above step again, dry the pellet, and dissolve it in required amount of nuclease free water (*see* **Note 11**).

3.2 Measurement of Quantity and Quality of the Extracted Plant DNA

1. Nanodrop is used to measure the DNA quantity and quality parameter. Generally, DNA is estimated to be pure if the ratios A260/A280 and A230/A280 are around 1.9.
2. Higher value indicates the presence of contaminants (*see* **Note 13**).
3. Furthermore, agarose gel electrophoresis can be performed for visualization. Poor quality DNA can be observed on the gel as smeared bands.

4 Notes

1. The selection of tissue and the storage technique are essential for the effectiveness of DNA extractions. Although extracts from freshly obtained tissue that are frozen in liquid nitrogen usually give the best results, pure and high molecular weight DNA can be extracted from tissues desiccated in silica gel [5].
2. Use high quality and fresh reagents especially CTAB buffer and isopropanol, which is essential for efficient DNA isolation [1].
3. The CTAB buffer should be at high concentration with added 2-β-mercaptoethanol to help in the lysing cells and denaturing proteins [5].
4. Maintain the temperature during the extraction process, especially ensuring that isopropanol is ice-cold for effective DNA precipitation.
5. Tissues that are partially crushed or dissolved prevent the high molecular weight DNA from being extracted. Mechanical maceration (using liquid nitrogen grinding to keep fresh tissues frozen) produces high quality DNA compared to chemical maceration due to its reduced phenolic compound reaction time or nuclease activity [5].
6. The chloroform:isoamylalcohol volume should be same. The phase in which the chemicals suspend in the lysate is determined by their solubility. Upper aqueous phase contains RNA, DNA, and polysaccharides whereas, lower phase contains plant debris, lipids, and proteins. Phenolic compounds are present in both phases [1].
7. The final stage of cleaning eliminates all traces of ethanol, which is essential as it inhibits the PCR amplification. Prolonged and high forced centrifugation or excessive drying of DNA pellet can result in low molecular weight or damaged DNA, which will prevent the pellet from dissolving in the elution buffer [5].
8. Low DNA yield could be due to inadequate cell lysis or incomplete precipitation of DNA. Sample should be well homogenized and increase incubation time with CTAB buffer, if necessary [3].
9. If DNA is degraded or sheared, it might be due to prolonged exposure to high temperatures or due to excessive mechanical forces while grinding [5].
10. Contamination with polysaccharides and polyphenols can interfere with DNA extraction and downstream applications. Modify the CTAB buffer by increasing the concentration of 2-β-mercaptoethanol, CTAB and use higher concentrations of

sodium chloride and potassium acetate to help remove these contaminants [1].

11. Poor DNA solubility after precipitation can be improved by adjusting pH of resuspension buffer or by using warm water bath to gently dissolve the pellet [3].
12. Contaminants like proteins and secondary metabolites can inhibit the downstream application. So, ensure proper pellet washing with 70% ethanol and consider additional purification steps if necessary [3].
13. Highly viscous DNA solution indicates the presence of high molecular weight DNA or contaminants. To reduce the viscosity, limit the mechanical shearing by pipetting gently and avoid vortexing. If viscosity is due to contaminants, additional purification steps may be necessary [3].
14. Sometimes, DNA may not pellet properly during precipitation step. To improve DNA pelleting, ensure that isopropanol is ice-cold to promote DNA precipitation, increase the centrifugation time or speed to ensure that DNA forms a solid pellet [2].

References

1. Islam W, Zaynab M, Qasim M, Wu Z (2017) Plant-virus interactions: disease resistance in focus. Hosts Viruses 4(1):5–20
2. Saini P, Aneja B, Yadav NR, Yadav RC (2013) Isolation of genomic DNA from leaf samples of Indian mustard without liquid nitrogen for use in molecular marker analysis. Crop Improv 40(1):30–33
3. Turaki AA, Ahmad B, Magaji UF, Yusuf BA, Hamza AB (2017) Optimised Cetyltrimethylammonium bromide (CTAB) DNA extraction method of plant leaf with high polysaccharide and polyphenolic compounds for downstream reliable molecular analyses. Afr J Biotechnol 16(24):1354–1365
4. Allen GC, Flores-Vergara MA, Krasynanski S, Kumar S, Thompson WF (2006) Natl Lib Med 1(5):2320–2325
5. Varma A, Padh H, Shrivastava N (2007) Plant genomic DNA isolation: an art or a science. Biotechnol J Healthc Nutr Technol 2:386–392
6. Pratyusha S (2022) Phenolic compounds in the plant development and defense: an overview. In: Hasanuzzaman M, Nahar K (eds) Plant stress physiology: perspectives in agriculture. IntechOpen, London
7. Sharma AD, Gill PK, Singh P (2002) DNA isolation from dry and fresh samples of polysaccharide-rich plants. Plant Mol Biol Report 20:415
8. Schenk JJ, Becklund LE, Carey SJ, Fabre PP (2023) What is modified CTAB protocol? Characterizing modifications to the CTAB DNA extraction protocol. Appl Plant Sci 11(3): e11517

Chapter 14

Extraction of High-Quality Genomic RNA Using Modified Trizol Methods

Sumiah Wani, Insha Mehraj, Nulevino Iralu, Dasari Meghanath, and Aflaq Hamid

Abstract

Invitrogen's TRIzol Reagent is a simple, all-in-one solution for isolating total RNA from a variety of biological materials. TRIzol is designed for PCR analysis and successfully disintegrates cells while preserving RNA integrity during homogenization.

Its distinct phenol-guanidine isothiocyanate composition aids in phase separation using chloroform, allowing RNA to be extracted into the aqueous phase. Isopropanol precipitation and nuclease-free water resuspension produce pure RNA appropriate for downstream uses. It simplifies the RNA isolation process as well as protects RNA from destruction and being compatible with several sample types makes it easy to use.

Key words Viral disease diagnostics, Plant virology, RNA viruses, Extraction, TRIzol, High quality genomic RNA

1 Introduction

RNA extraction is fundamental in plant virology for several key reasons. It allows researchers to isolate viral RNA from plant tissues, which is crucial for detecting and identifying plant viruses. Since many plant viruses have RNA genomes, extracting viral RNA enables accurate and sensitive diagnosis of infections [1]. RNA extraction also facilitates the study of viral gene expression, helping scientists understand how viruses replicate and interact with host plants. This knowledge is essential for developing targeted control measures and breeding virus-resistant plant varieties. Additionally, RNA extraction supports research into viral diversity and evolution, contributing to better management strategies and improved crop protection [2]. Overall, RNA extraction is critical for advancing our understanding of plant virology and enhancing agricultural practices. In plant virology lab, RNA extraction is performed frequently. High quality RNA extraction from the plant sample is critical to

Aflaq Hamid et al. (eds.), *Detection of Plant Viruses: Advanced Techniques*, Springer Protocols Handbooks,
https://doi.org/10.1007/978-1-0716-4390-7_14,

succeed in different experiments like cDNA synthesis, reverse transcription-polymerase chain reaction (RT-PCR), northern hybridization, cDNA library construction, gene amplification, and microarray analysis [3]. The overall efficiency of a virus detection method is dependent on the first step itself, i.e., the extraction process, which necessitates the proper understanding of the protocol and expertise skill of the technique. The method follows a series of steps sequentially with proper handling of chemicals in aseptic laboratory conditions. TRIzol is an acid guanidinium-phenol-based reagent ideally designed for the extraction of RNA, and simultaneously extracts DNA as well as protein. The guanidinium salt serves as a chaotropic agent to denature proteins and the phenol is an organic compound also used to extract nucleic acids and proteins [4–7]. TRIzol is considered as a gold standard for RNA purification. It is an all-in-one reagent that offers excellent capability to lyse a variety of complex samples [8]. It is easily scalable for accommodating very large inputs, recovers ample RNA yield, and is also an effective inactivator of RNases, preventing degradation of samples during purification [9].

2 Materials [10]

1. Plant material that is suspected of viral infection, based on symptoms.
2. TRIzol reagent from any manufacturer like TRIzol by Invitrogen or TRI Reagent from Sigma-Aldrich.
3. Chloroform.
4. 75% ethanol.
5. Nuclease-free water/RNase-free water.
6. Isopropanol.

3 Methods

3.1 Sampling

Collected plant sample should be immediately stored in liquid nitrogen. For long-term storage, the plant tissues can be stored in −80 °C.

3.2 Grinding

To obtain high yield during the extraction of genetic material, grinding of the plant sample is a critical element. Maintaining the samples at low temperature as long as possible during grinding to avoid degradation and grinding the tissue as finely as possible is highly essential. This step must be carried out in liquid nitrogen in a sterilized mortar and pestle, crushing the samples to fine powder.

3.3 Extraction

1. Grind 100 mg of plant samples into powder in liquid nitrogen using prechilled mortar and pestle (*see* **Note 1**).
2. Transfer the crushed plant sample to microtubes and add 1 mL of TRIzol reagent (*see* **Note 2**).
3. Add chloroform 200 uL per 1 ml of TRIzol reagent (*see* **Note 3**).
4. Mix by shaking and incubate at room temp for 2–3 min.
5. Centrifuge the sample at 12000 g for 15 min at 4 °C.
6. Transfer supernatant (aqueous phase) to new tube and add 500 uL isopropanol (*see* **Note 6**).
7. Incubate for 2 h at −80 °C or overnight at −20 °C.
8. Centrifuge for 10 min at 12000 g at 4 °C.
9. Discard the supernatant and wash pellet with 500 uL of 75% ethanol and centrifuge for 5 min at 7500 g at 4 °C. Repeat this step two to three times.
10. Discard the supernatant and dry pellet under laminar air flow hood, for not more than 10–15 min.
11. Resuspend the cells in RNase-free water and store in −80 °C until further use.

4 Notes

1. Always read protocol before carrying out the experiment. When doing it for the first time, be under the supervision of a senior or expert. Avoid walking, talking, eating and other activities while carrying out the experiment.
2. Working in a molecular laboratory, especially while dealing with viruses necessitates quicker mode of action i.e., speed. Aseptic conditions are to be maintained at every step to avoid any contamination which can lead to nucleic acid degradation.
3. Store TRIzol away from light as it is photosensitive. Reagents should be used as per manufacturer's protocols and modified only when required as per laboratory conditions. Use the appropriate amount of reagents and chemicals; do not over or under use.
4. Pipetting error should be reduced to as low as possible to avoid false results.

References

1. Mehetre GT, Leo VV, Singh G, Sorokan A, Maksimov I, Yadav MK, Upadhyaya K, Hashem A, Alsaleh AN, Dawoud TM, Almaary KS (2021) Current developments and challenges in plant viral diagnostics: a systematic review. Viruses 13(3):412

2. Rubio L, Galipienso L, Ferriol I (2020) Detection of plant viruses and disease management: relevance of genetic diversity and evolution. Front Plant Sci 11:1092
3. Sasi S, Krishnan S, Kodackattumannil P, Shamisi AA, Aldarmaki M, Lekshmi G, Kottackal M, Amiri KM (2023) DNA-free high-quality RNA extraction from 39 difficult-to-extract plant species (representing seasonal tissues and tissue types) of 32 families, and its validation for downstream molecular applications. Plant Methods 19(1):84
4. Thatcher SA (2018) Nucleic acid isolation. In: Principles and applications of molecular diagnostics. Elsevier, pp 35–46
5. Chomczynski P, Wilfinger W, Kennedy A, Rymaszewski M, Mackey K (2010) RNAzol® RT: a new single-step method for isolation of RNA. Nat Methods 7(12):4–5
6. Chomczynski P, Sacchi N (1987) Single step method of RNA isolation by acid guanidine thiocyanate-phenol-chloroform extraction. Anal Biochem 162:156–159
7. Brawerman G, Mendecki J, Sy L (1972) A procedure for the isolation of mammalian messenger ribonucleic acid. Biochemistry 11(4): 637–641
8. Martins R, Queiroz JA, Sousa F (2014) Ribonucleic acid purification. J Chromatogr 1355: 1–14
9. Beintema JJ, van der Laan JM (1986) Comparsion of the structure of turtle pancreatic ribonuclease with those of mammalian ribonucleases. FEBS Lett 194(2):338–342
10. Anonymous (2023) TRIzol reagent literature. Invitrogen Life Technologies

Chapter 15

Extraction of High-Quality Genomic RNA Using Modified CTAB Method

Sumiah Wani, Nulevino Iralu, Dasari Meghanath, and Aflaq Hamid

Abstract

RNA extraction from plants, particularly perennials, is sometimes difficult due to the presence of interfering chemicals such as polysaccharides and polyphenolics. The CTAB method is widely used for separating nucleic acids from plant material. However, this method necessitates customization to certain plant species and tissues. The modified CTAB approach applies to a wide variety of woody and herbaceous plants, but this approach includes higher concentrations of β-mercaptoethanol. This technique produced high-quality RNA with low degradation. The extracted RNA is useful for further transcriptional analysis, such as RT-qPCR. The modified CTAB method presented here is a reliable and effective method for isolating RNA from woody plants, enabling molecular genetic research of these economically important species.

Key words Virus, RNA, Extraction, Plant, CTAB

1 Introduction

High-quality RNA is required for molecular genetic investigations; however, its extraction from plant tissues can be difficult due to the presence of secondary compounds such as polysaccharides and polyphenols [1]. These substances can interfere with RNA extraction and degrade the RNA, making it unsuitable for subsequent uses. Phenolic chemicals, in particular, can oxidize into quinones, which bind to RNA and prevent it from being used in procedures such as reverse transcription and cDNA library creation [2]. The majority of woody dicotyledonous plant tissues contain secondary metabolites such as polysaccharides and polyphenols [3]. Commercial RNA isolation kits frequently employ acid guanidinium thiocyanate-phenol-chloroform extraction, which may yield low-quality RNA [4]. To address these obstacles, the CTAB (cetyl trimethylammonium bromide) approach is frequently employed to extract nucleic acids from polysaccharide-rich plant tissues [5–7]. CTAB extraction buffers commonly include CTAB, PVP,

Aflaq Hamid et al. (eds.), *Detection of Plant Viruses: Advanced Techniques*, Springer Protocols Handbooks,
https://doi.org/10.1007/978-1-0716-4390-7_15,

sodium chloride, and β-mercaptoethanol, which all contribute to the extraction process. While the CTAB approach is extensively used, it may need to be adapted to specific plant species and the extraction processes should be carefully optimized [8]. Cetyl trimethyl ammonium bromide (CTAB), sometimes called as cetrimonium bromide, is a quaternary ammonium salt with surface-active and antiseptic properties. It was first mentioned in the chemical literature in 1936 by G.S. Hartley, B. Collie, and C.S. Samis at University College London. Shelton et al. reported a synthesis of CTAB from cetyl (hexadecyl) bromide and trimethylamine in 1946 at the William S. Merrell Company (Cincinnati) [9]. In addition to its application as a topical antiseptic in chemical industry, it is popularly used in molecular biology for DNA and RNA extraction processes [10, 11]. It works as an ionic (cationic) detergent that disrupts cellular membranes and separates nucleic acids from polysaccharides and CTAB-RNA complexes.

2 Materials

1. CTAB buffer.
2. Chloroform: Isoamylalcohol (24:1).
3. β-Mercaptoethanol.

3 Methods

1. Preheat 100 mL of buffer at 65 °C in a water bath. Add 700 uL of β-mercaptoethanol (*see* **Note 1**).
2. Crush 100 mg of plant tissues in liquid nitrogen and transfer sample to microtube and add 1 mL of CTAB buffer to it (*see* **Note 2**).
3. Incubate at 65 °C for 10 min and then cool at room temperature.
4. Centrifuge at 8000 g for 2 min.
5. Transfer supernatant to new tube and add equal volume of chloroform:isoamylalcohol (24:1).
6. Mix by shaking and centrifuge for 10–20 min at 11000 g.
7. Collect supernatant and again add chloroform:isoamylalcohol (24:1) in equal volume and centrifuge again at 11000 g for 10–20 min.
8. Take supernatant again in new tube and add $LiCl_2$ in equal ratio (1:1).
9. Incubate the mixture at −20 °C overnight or at −80 °C for 1–2 h.

10. Centrifuge at 21000 g for 20 min at 4 °C and retain the pellet.
11. Wash pellet with 80% ethanol and centrifuge at 14000 g for 10 min.
12. Air dry under LAF hood for 10–15 min and resuspend in required amount of nuclease-free water and store at −80 °C for further use (*see* **Note 3**).

4 Notes

1. Use high-quality, certified reagents like nuclease-free or DEPC-treated water. Freshly prepared buffers should be used and should be made ahead of time.
2. To prevent RNase activity, precool all the tools that are to be used in the RNA extraction process. To prevent contamination, use DEPC treated mortar/pestle and plastic wares, and for each sample use separate mortar/pestle should be used.
3. To prevent nucleic acid degradation, maintain aseptic conditions throughout the process. Wear gloves while handling RNA and change them frequently.

References

1. Gehrig HH, Winter K, Cushman J, Borland A, Taybi T (2000) An improved RNA isolation method for succulent plant species rich in polyphenoles and polysaccharides. Plant Mol Biol Rep 18:369–376
2. Oulakakis KA, Roubelakis-Angelakis KA, Kanellis AK (1996) Isolation of functional RNA from grapevine tissues poor in nucleic acid content. Am J Enol Vitic 47:181–185
3. MacKenzie DJ, McLean MA, Mukerji S, Green M (1997) Improved RNA extraction from woody plants for the detection of viral pathogens by reverse transcription-polymerase chain reaction. Plant Dis 81(2):222–226
4. Gambino G, Perrone I, Gribaudo I (2008) A rapid and effective method for RNA extraction from different tissues of grapevine and other woody plants. Phytochem Anal 19:520–525
5. White EJ, Venter M, Hiten NF, Burger JT (2008) Modified cetyltrimethylammonium bromide method improves robustness and versatility: the benchmark for plant RNA extraction. Biotechnol J 3:1424–1428
6. Vennapusa AR, Somayanda IM, Doherty CJ, Jagadish SVK (2020) A universal method for high–quality RNA extraction from plant tissues rich in starch, proteins and fiber. Sci Rep 10: 16887
7. Yan WJ, Pendi FH, Hussain H (2022) Improved CTAB method for RNA extraction of thick waxy leaf tissues from sago palm (*Metroxylon sagu* Rottb). Chem Biol Technol Agric 9:63
8. Kiss T, Karácsony Z, Gomba-Tóth A, Szabadi KL, Spitzmüller Z, Hegyi-Kaló J, Cels T, Otto M, Golen R, Hegyi ÁI, Geml J (2024) A modified CTAB method for the extraction of high-quality RNA from mono-and dicotyledonous plants rich in secondary metabolites. Plant Methods 20(1):62
9. Shelton D, Stranne M, Mikkelsen L, Pakseresht N, Welham T, Hiraka H, Tabata S, Sato S, Paquette S, Wang TL, Martin C (2012) Transcription factors of lotus: regulation of isoflavonoid biosynthesis requires coordinated changes in transcription factor activity. Plant Physiol 159(2):531–547
10. Anonymous (2023) Cetyltrimethylammonium bromide molecule of the week archive. American Chemical Society
11. Wani S, Shah MD, Padder BA, Nabi SU, Hamid A, Mir MM, Ara S (2024) Occurrence of cherry virus a (CVA) infecting stone fruits in Kashmir province of northwestern Himalayan region of India. Indian Phytopathol 77(1): 139–145

Chapter 16

Extraction of High-Quality Genomic RNA Using Modified Kit Method

Sumiah Wani, Nulevino Iralu, Dasari Meghanath, and Aflaq Hamid

Abstract

RNA extraction kits are a convenient and fast way to isolate RNA from a variety of biological materials. These kits typically use silica-based spin columns or magnetic beads to bind RNA molecules while also eliminating impurities like proteins and DNA. Cell lysis, RNA binding to the solid phase, contamination removal, and pure RNA elution are all common processes in the extraction process. RNA extraction kits are commonly used in molecular biology research for purposes such as gene expression analysis, microarray investigations, and RNA sequencing. They have numerous benefits, including ease of use, repeatability, and high-quality RNA yields.

Key words Plant, Virus, RNA, Extraction, Kits

1 Introduction

RNA extraction from virus-infected plant tissues is an important step in understanding viral infections and creating diagnostic tools. Viruses are obligate intracellular parasites that reproduce within host cells and produce viral RNA or DNA. The extraction and analysis of viral nucleic acid reveals important information on the virus's genetic makeup and the host's response to infection. However, extracting high-quality RNA from virus-infected plant tissues presents several obstacles since plants frequently contain significant quantities of polysaccharides, which can interfere with RNA extraction by adhering to the RNA and hindering isolation [1, 2]. Plant tissues contain secondary metabolites, such as phenolics and alkaloids, which can disrupt extraction by blocking enzymes or binding to RNA. Furthermore, some viruses generate proteases, which can damage RNA during the extraction process [3–5]. Commercial plant RNA extraction kits from companies such as Qiagen and Thermofisher Scientific are popular. They use a silica-based

Aflaq Hamid et al. (eds.), *Detection of Plant Viruses: Advanced Techniques*, Springer Protocols Handbooks,
https://doi.org/10.1007/978-1-0716-4390-7_16,

membrane to bind RNA and are built to withstand high polysaccharide and secondary metabolite content [6].

2 Materials

1. RNA Plus Extraction Kit (Invitrogen, Thermofisher Scientific, CA, USA).
2. TriZol.
3. Liquid Nitrogen.

3 Methods

1. Leaf samples were crushed in liquid nitrogen, and 1 mL of Trizol reagent was added. Samples were homogenized by pipetting up and down several times (*see* **Note 1**).
2. 200 μL of chloroform was added to each tube after incubation of 5 min at room temperature (*see* **Note 2**).
3. Each sample was centrifuged at 1200 rpm in a refrigerated centrifuge for 15 min.
4. Aqueous phase obtained was transferred to another tube and diethyl pyrocarbonate (DEPC)-treated ethanol was added to each tube in the ratio of 1:1 followed by vortexing (*see* **Note 3**).
5. About 700 μL from each sample was transferred to spin cartridge followed by centrifugation for 15 s and the step was repeated twice.
6. Harvest cells:
   (a) Place the frozen cell pellet into an appropriately sized RNase-free tube.
   (b) Add the appropriate volume of Lysis Buffer with 2-mercaptoethanol to your sample (Prepare fresh Lysis Buffer).
   (c) Vortex at high speed until the cell pellet is completely dispersed and the cells appear lysed.
7. Homogenize Cells:

   Centrifuge the homogenate at approximately 2600× *g* for 5 min, then transfer the supernatant to a clean RNase-free tube.
8. Wash RNA:
   (a) Add 700 μL of Wash buffer I to the spin cartridge.
   (b) Centrifuge at 12,000× *g* for 15 s at room temperature.

(c) Discard the flow-through and the collection tube. Place the spin cartridge into a new collection tube.

(d) Add 500 μL Wash Buffer II with ethanol to the spin cartridge.

(e) Centrifuge at 12,000× *g* for 15 s at room temperature.

(f) Discard the flow-through and reinsert the spin cartridge in the same collection tube.

(g) Repeat steps d-f one more time.

9. Elute RNA:

(a) Centrifuge the spin cartridge with collection tube at 12,000× *g* for 1 min at room temperature.

(b) Discard the collection tube and insert the spin cartridge into a recovery tube.

(c) Add 30 μL to 3 × 100 μL RNase-free waste to the center of the spin cartridge.

(d) Incubate at room temperature for 1 min.

(e) Centrifuge at 12,000× *g* for 2 min at room temperature.

Note: Collect all eluates into the same tube when performing serial elution.

10. Analyze RNA yield and quality using UV absorbance at 260 nm.

11. Store purified RNA at −80 °C for long-term storage.

4 Notes

1. Use clean gloves while handling reagents, reaction vessels or samples. Replace your gloves frequently, especially after touching common surfaces.

2. Prevent RNase contamination in pipette and rinse the pipette with RNase decontamination solution.

3. Sterilizing labware makes RNases ineffective. Properly autoclave tips, tubes, and solutions and bake glassware at 300 °C for 4 h.

(a) Use DEPC treated water for plasticware, tubes, and most solutions. Rinse treated labware twice with distilled water.

(b) Use RNase decontamination reagents on lab equipment, surfaces, and pipettes. These chemicals can be poured, sprayed, or rubbed over surfaces before starting work.

References

1. Li Z, Trick HN (2005) Rapid method for high-quality RNA isolation from seed endosperm containing high levels of starch. BioTechniques 38:872–876. https://doi.org/10.2144/05386BM05
2. Wang G, Wang G, Zhang X, Wang F, Song R (2012) Isolation of high quality RNA from cereal seeds containing high levels of starch. Phytochem Anal 23:159–163. https://doi.org/10.1002/pca.1337
3. White EJ, Venter M, Hiten NF, Burger JT (2008) Modified cetyltrimethylammonium bromide method improves robustness and versatility: the benchmark for plant RNA extraction. Biotechnol J 3:1424–1428. https://doi.org/10.1002/biot.200800207
4. Rezadoost MH, Kordrostami M, Kumleh HH (2016) An efficient protocol for isolation of inhibitor-free nucleic acids even from recalcitrant plants. 3 Biotech 6:61. https://doi.org/10.1007/s13205-016-0375-0
5. Singh G, Kumar S, Singh P (2003) A quick method to isolate RNA from wheat and other carbohydrate-rich seeds. Plant Mol Biol Report 21:93a. https://doi.org/10.1007/BF02773401
6. Mahiya F (2017) Gene expression profiling in phaseolus vulgaris-colletotrichum lindemuthianum pathosystem, Doctoral dissertation, SKUAST Kashmir

Chapter 17

Gel Electrophoresis for Visualization of Nucleic Acid

Andleeb Javaid, Sumiah Wani, Nulevino Iralu, Gowhar Ali, and Aflaq Hamid

Abstract

Gel electrophoresis is a procedure used in molecular biology to identify and separate molecules (such as DNA, RNA, protein complexes) with varying conformations, charges, or size. Loading RNA and DNA onto the gels enables the assessment of fragment sizes by separating the RNA and DNA fragments. This is achieved by the use of charge within the particles. Distinct bands at set sizes are formed when these fragments separate.

Key words Gel, RNA, DNA, Bands, Electrophoresis

1 Introduction

In the past few years, forensic, medical applications, as well as research and infectious organism identification and characterization have all made use of molecular biology techniques such as polymerase chain reaction. In the field of virology, it has been demonstrated that using PCR approach provides the benefits of high reproducibility and high sensitivity in viral genomic detection and strains characterization, but the sensitivity in DNA fragment detection is also associated with the electrophoresis matrix used for development of PCR product [1]. Biomolecules frequently have association with electrical charges, either negative or positive. Due to electrostatic attraction, charged biomolecules gravitate toward the opposite charged electrode when placed in an electric field. Gel electrophoresis is a procedure used in molecular biology to identify and separate molecules (such as DNA, RNA, protein complexes) with varying conformations, charges, or size. Various factors influence the relative mobility of individual molecules, temperature, porosity, and viscosity of the matrix through which the molecule migrates, as well as the net charge, charge/mass ratio, and

Aflaq Hamid et al. (eds.), *Detection of Plant Viruses: Advanced Techniques*, Springer Protocols Handbooks,
https://doi.org/10.1007/978-1-0716-4390-7_17,

molecular shape are the most crucial factors. This method allows for the very high resolution separation of complex mixtures [2].

The separation of these molecules is achieved by placing them in a gel made up of small pores and setting an electric field across the gel. These molecules move through the gel at various speeds and directions depending on their size and charge. Standard method of electrophoresis is through agarose or polyacrylamide gels as both these gels are porous and are used to identify, separate, and purify nucleic acids. Separation is therefore dependent on both the charge and the size of the molecule because the gels are porous and the size of the pores in relation to the molecule decides whether the molecule will enter the pore and be retarded or will bypass it. When compared to other types of media, gels provide better and sharper sample resolution [3]. Gel slab is created with either specific concentration of polyacrylamide concentration or agarose. The sample passes through the matrix formed by the gel [4]. The matrix which separates the sample is a cross-polymer [4]. The tightness of "matrixes" mesh is determined by the concentration of agarose, the length of fragment (kilobase pairs) visualized through the slab is determined by the various concentrations [5]. The higher percentage of agarose (2%) will show a lower KB (0.2–1 KB), whereas lower percentage of agarose (0.7%) will show a higher KB (5–10 KB [6]. Rather than employing agarose at a higher percentage, polyacrylamide is used to visualize smaller KB pairs [7].

To make the DNA/RNA in the gel more visible, ethidium bromide is added to the gel. It has been demonstrated that virus DNA/RNA can be visualized through gel electrophoresis [8, 9]. Initially, DNA quality was determined using ultracentrifuges [10]. This technique used machinery involved with the centrifuge, which required constant high speed motion that frequently led to equipment failure. Piet Borst and Cees Aaij came up with the concept to attempt and visualize mitochondrial DNA using gel electrophoresis during one of these breakdowns. Cold DNA would not be easily seen using electrophoresis, however, hot DNA would be easily seen. It was visualized by adding ethidium bromide, which made consistent visualization possible [10]. The toxicity of ethidium bromide as a dye for gel electrophoresis is one of the concern with its use. The chemical is both mutagenic and carcinogenic. As a result, different staining techniques are being examined [11]. Instead of ethidium bromide, dyes like GelRed and GelGreen can be employed for the process of staining [12]. The buffer in the electrophoresis chamber and the power supply attached to it are responsible for the visibility [13, 14]. Tris/Borate/EDTA (TBE) or Tris/Acetate/EDTA (TAE) are the buffers that are used in electrophoresis. By conducting an electrical charge through the buffer, the electrical charge of samples is attracted to the electrical current and directed toward the positive terminal corresponding to their weight. Then depending on their

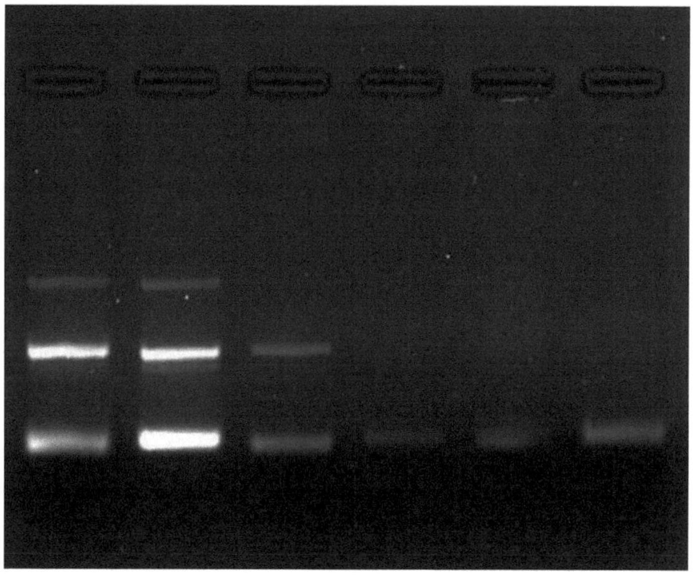

Fig. 1 One percent gel visualized under Gel Doc system (Chapter 17)

size, the sample will coalesce into bands along the gel at various lengths as shown in Fig. 1 [15, 16]. Ladder markers can be used to identify these bands and determine the KB size of the band [10]. Identification of KB size in the sample is possible because of known KB size of ladder markers.

1.1 Principle of Electrophoresis

A mixture of electrically charged biomolecules will freely migrate toward the electrode with the opposite charge if they are exposed to an electric field of strength E. However, depending on the physical properties of the molecule and the system of experiment, various molecules will move at highly different and individual rates. Velocity (v) of movement of a charged molecule in an electric field depends on variables denoted by

$$v = E.q/f$$

where q is the net charge on the molecule and f is the frictional coefficient [17]. The frictional coefficient (f) depends on its degree of compactness, mass of its molecule, porosity of the matrix in which experiment is conducted and buffer velocity, and it represents the frictional resistance to mobility. The number of positive and negative charges within the molecule determines its net charge. Proteins carry charges due to the presence of side chains of amino acids or groups resulting from post-translational changes like, phosphorylation, acylation, or deamidation. Since each nucleotide in DNA has single negative charge conferred by a phosphate group [18], DNA exhibits uniform charge distribution. The above equation in general indicates that the molecules will generally move quickly as their net charge increases, the electric field strengthens, and f (function of mase/shape) decreases. Similar mass/shape molecules may differ widely in net charge, although molecules with similar net charge separate because of variation in their frictional coefficients. As a result, electrophoresis frequently gives very high-resolution separation. Nowadays, agarose-based polysaccharide gel matrix is the most commonly used. This polymer is the repeating unit of disaccharide called agarobiose, which is made up of galactose and 3,6-anhydrogalactose. When compared with starch, agarose has more uniform degree of porosity, which may be adjusted by changing the initial concentration of suspension (high concentration gives smaller pores whereas low concentration produces larger pores). The separation of DNA molecules is the primary application for this gel, while it can also be used for other electrophoretic processes involving protein materials, such as immunoelectrophoresis [8]. Based on mobility in agarose gels, DNA molecular masses can be precisely determined due to the uniform charge distribution in nucleic acids [19].

2 Materials Required

2.1 Reagents

1. 1× TAE.
2. DNA loading buffer.
3. Deionized water.
4. Agarose stock solution for DNA gels.

2.2 Equipments (Fig. 2)

1. Laboratory wipes.
2. Beaker.
3. Gel plates.
4. Gloves.
5. Combs.
6. Ethanol, glass cleaner, soap.
7. Electrophoresis apparatus (Fig. 3).
8. Pipette.
9. Microwave.
10. Imaging equipment.

3 Method

1. Gel plates should be cleaned with soap and water to remove any dried acrylamide or particulate.
2. Treat the plate and comb with ethanol using the tissue paper. And allow it to dry.

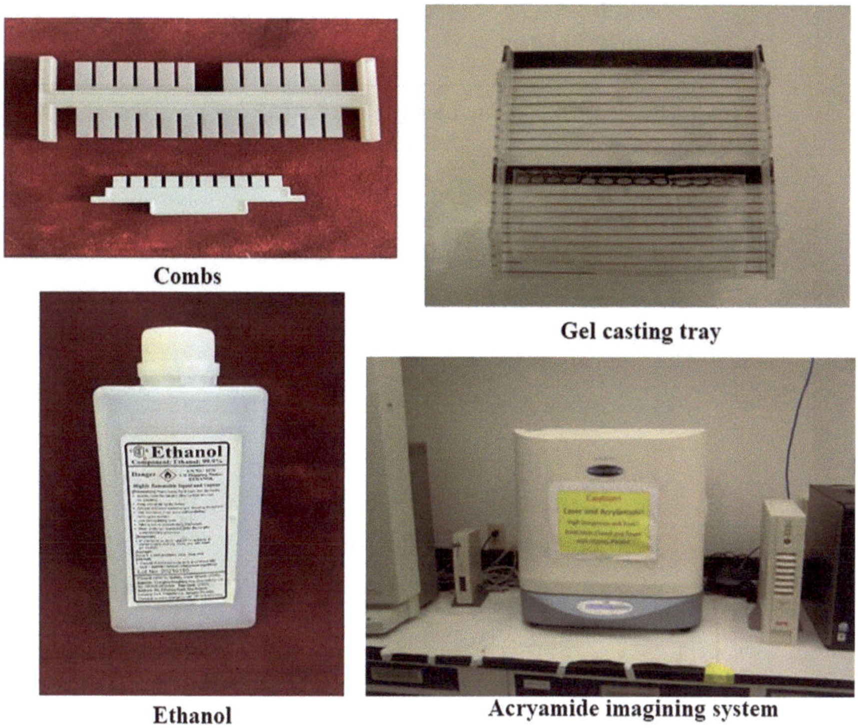

Fig. 2 Equipments required for gel electrophoresis (Chapter 17)

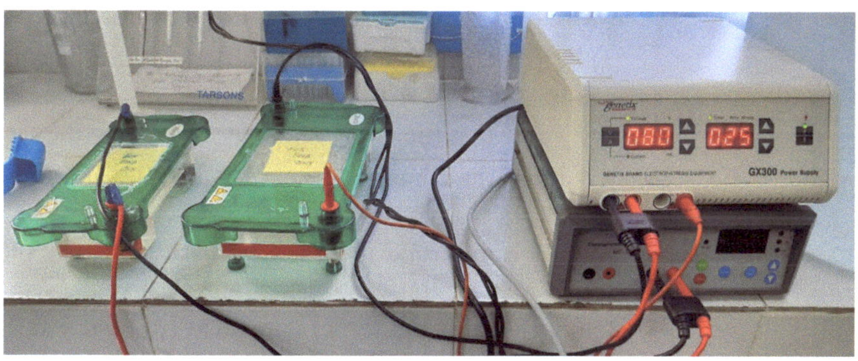

Fig. 3 Gel electrophoresis unit (Chapter 17)

3. Use Table 1 to create the desired gel solution based on the visualization desired.
4. Stir the gel solution.
5. Heat in the microwave until the agarose dissolves entirely. After every 30 s, stop the microwave and stir the mixture. Be careful while you stir. When stirring, wear safety gloves and lab coat because the solution may boil over and cause serious burns.
6. Cool the agarose to 60 °C.

Table 1
Agarose solution

% Agarose	Agarose (g)	1× TAE (mL)	Visualization size (bp)
0.5	0.5	100	1000–30,000
0.7	0.7	100	800–12,000
1	1.0	100	500–10,000
1.2	1.2	100	400–7000
1.5	1.5	100	200–3000
2	2.0	100	50–2000

7. Add 50 μg of ethidium bromide to molten agarose to help visualization.
8. Prepare plates for gel solution.
   (a) Place the spacer on the plate's inside edge.
   (b) Put the second plate on the top of long plate, agent side down.
   (c) Use binder clips to secure.
   (d) Tightly fasten the comb.
   (e) Secure plates with tape along the edge.
9. Before the gel polymerizes, quickly fill the assembled plates.
   (a) Tip the plate so that the solution falls to the edge closest to you, then lay it horizontally.
   (b) Verify that the entire region is covered by the solution.
10. Make sure the complete absence of bubbles and firmly clamp the comb between the two plates.
11. Wait for some time in order to make sure that the comb-formed well separations are accurate.
12. Remove all clips and other items used to hold plates in place after the gel has solidified.
13. Take the comb out and wash the comb region.
14. Fit the gel apparatus with the gel plates.
15. Add 1× TAE buffer.
16. Remove the lid and Flush wells with 1× TAE.
17. Add sample to the wells.
18. Allow to run for min at constant Watts (80 W) based on the size of samples.
19. Turn off the power supply.
20. Take the gel out of the spacers and apparatus.

21. Use spatula to carefully separate the gel from the apparatus so that the gel is retained on the smaller plate.
22. Visualize the gel in gel imaging system.

3.1 Applications of Gel Electrophoresis

The agarose gel electrophoresis technique was widely used to examine the DNA cleavage effectiveness of small molecules as well as a useful method to examine different ways in which small molecules can attach to supercoiled DNA [20]. This was primarily due to the significance of DNA cleavage in drug design. The predominant forms of naturally produced plasmid DNA are closed circle supercoiled (SC), with trace amounts of nicked circular form (NC) and linear forms. Using the agarose gel electrophoresis method, the relaxation of supercoiled (SC) pUC19 into linear conformation (LC) and nicked circular (NC) can be utilized to measure the relative cleavage effectiveness of complexes. It is a useful technique to examine different ways that small molecules bind to supercoiled DNA [21]. Small molecules can intercalate into plasmid DNA to loosen or cleave the SC DNA form, which lowers its mobility rate and can be visualized separately by agarose gel electrophoresis method while the SC form of plasmid DNA is not significantly affected by simple electrostatic interaction of small molecules to DNA, hence the mobility of supercoiled DNA remains unchanged [22] (Fig. 4).

4 Notes

1. Ensure your DNA samples are properly purified and quantified before loading them onto the gel. Contaminants or impurities can affect the results [1].
2. Use high-quality reagents and nuclease-free labware deep and narrow wells to avoid the formation of faint bands [24].
3. Load appropriate sample volume and mix the gel components thoroughly so that there is no formation of smeared bands [24].
4. Load samples carefully into wells, ensuring consistent volume and avoiding air bubbles [25].
5. Always include ladder for size reference [26].
6. Handle ethidium bromide and UV light with caution during staining and visualization, because ethidium bromide is a known mutagen [25].
7. Optimize voltage and run time based on fragment size and gel type [27].
8. Maintain a consistent temperature during electrophoresis. Fluctuations can impact the migration of samples [28].

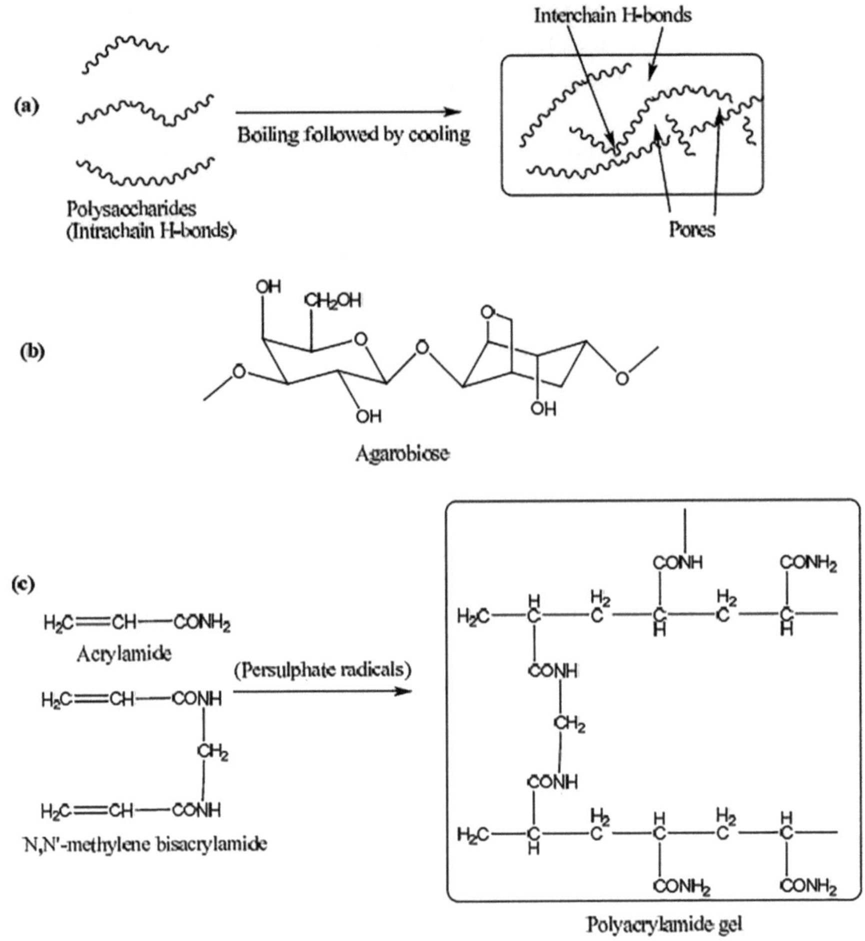

Fig. 4 Commonly used gels in electrophoresis of nucleic acids and proteins (Chapter 17)

9. Buffer should be carefully chosen based on your gel type and target molecule as it provides pH and electrical conductivity for gel in the electrophoresis chamber. It also affects the mobility and stability of molecules in the gel [25].
10. Avoid inappropriate stain, understaining, and overstaining in order to ensure the quality and accuracy of the image [27].
11. If some lanes are running unevenly from left to right, the gel is frequently blamed. While making the gel, ensure that the agarose gel is heated and mixed well, that the gel is cast on a level surface, and the comb is set straight across the gel. If the central lanes moves faster than the outside lane, the voltage is usually too high. When a random lane flows erratically, it is usually due to bubbles, inadequate heating damage to sample well, or inadequate mixing of molten agarose solution before gel casting [26].

References

1. Barril P, Nates S (2012) Gel electrophoresis principles and basics. IntechOpen, London
2. Sheehan D (2000) Physical biochemistry: principles and applications. Wiley, New York, pp 153–211
3. Mardali M, Sarraf-Mamoory R, Sadeghi B, Safarbali B (2016) Acrylamide route for the co-synthesis of tungsten carbide–cobalt nanopowders with additives. Ceram Int 42:9382–9386
4. Dolnik V (1997) Capillary zone electrophoresis of proteins. Electrophoresis 105(2): 2353–2361
5. Sajjadi SH, Ahmadzadeh H, Goharshad EK (2020) Enhanced electrophoretic separation of proteins by tethered SiO 2 nanoparticles in an SDS-polyacrylamide gel network. Analyst 145:415–423
6. Minden JS, Dowd SR, Meyer HE, Stuhler K (2009) Difference gel electrophoresis. Electrophoresis 30:5156–5161
7. Stuber CW, Lincoln SE, Wolff DW, Helentjaris T, Lander ES (1992) Identification of genetic factors contributing to heterosis in a hybrid from two elite maize inbred lines usng molecular markers. Genetics 132:823–839
8. Eckerskorn C, Lottspeich F (1989) Internal amino acid sequence analysis of protein separation by gel electrophoresis after tryptic digestion in polyacrylamide matrix. Chromotagraphia 28:92–94
9. Wani S, Shah MD, Padder BA, Nabi SU, Hamid A, Mir MM, Ara S (2024) Occurrence of Cherry Virus A (CVA) infecting stone fruits in Kashmir province of Northwestern Himalayan Region of India. Indian Phytopathol 77(1):139–145
10. Viovy JL, Duke T (2005) DNA electrophoresis in polymer solutions: ogston sieving, reptation and constraint release. Electrophoresis 14: 322–329
11. Johnson PH, Grossman LI (1977) Electrophoresis of DNA in agarose gels. Optimizing separations of conformational isomers of double and single stranded DNAs. Biochemistry 16:4217–4225
12. Petrov A, Tsa AS, Puglisi JD (2013) Analysis of RNA by analytical polyacrylamide gel electrophoresis. In: Lorsh J (ed) Methods in enzymology. Academic, pp 301–313
13. Rio DC, Ares M Jr, Hannon GJ, Nilsen TW (2010) Polyacrylamide gel electrophoresis of RNA. Cold Spring Harb Protoc. https://doi.org/10.1101/pub.prot5444
14. Thorne HV (1966) Electrophoretic separation of polyoma virus DNA from host cell DNA. Virology 29:234–239
15. Borst P (2005) Ethidium DNA agarose gel electrophoresis: how it started. IUBMB Life 57(11):745–747
16. Schmidt F, Schmid TJ, Riechers A, Haase S, Bosserhoff A, Heilmann J, Konig B (2010) DNA staining in agarose gels with ZN^{2+}-cyclen-pyrene. Nucleosides Nucleotides Nucleic Acids 29(10):687–703
17. Adamson NJ, Reynolds EC (1997) Rules relating electrophoretic mobility, charge and molecular size of peptides and proteins. J Chromatogr B 699:133–147
18. Tenover FC, Arbeit RD, Goering RV, Mickelson PA, Murray BE, Persing DH, Swaminathan B (1995) Interpreting chromosomal DNA restriction patterns produced by pulsed-field gel electrophoresis: criteria for bacterial strain typing. J Clin Microbio 33:2233–2239
19. Aebersold PB, Winans GA, Teel DJ, Milner GB, Utter FM (1987) Manual for starch gel electrophoresis: a method for detection of genetic variation. NOAA technical report NMFS 61
20. Li Y, Yang Z (2009) DNA binding affinity and antioxidative activity of copper(II) and zinc (II) complexes with a novel hesperetin Schiff base ligand. Inorg Chim Acta 362(13): 4823–4831
21. Chen Z-F (2011) Potential new inorganic antitumor agents from combining the anticancer traditional Chinese medicine (TCM) matrine with Ga(III), Au(III), Sn(IV) ions, and DNA binding studies. J Inorg Biochem 105(2): 171–180
22. Dowdle ME, Imboden SB, Park S, Ryder SP, Sheets MD (2017) Horizontal gel electrophoresis for enhanced detection of protein-RNA complexes. J Vis Exp 125:56031
23. Reddy PR (2012) Gel Electrophoresis: Nucleic Acids. IntechOpen
24. Thermo Fisher Scientific (n.d.) Nucleic Acid Gel Electrophoresis Troubleshooting
25. Sambrook J, Russell DW (2001). Molecular cloning: a laboratory manual. Cold spring harbor laboratory press
26. Brown TA (2006) Genomes. Garland science
27. Ausubel FM, Brent R, Kingston RE, Moore DD, Seidman JG, Smith JA (2002) Current protocols in molecular biology. John Wiley & Sons
28. Alberts B, Johnson A, Lewis J, Raff M, Roberts K, Walter P (2002) Molecular biology of the cell. Garland science

Chapter 18

Polyacrylamide Gel Electrophoresis for Visualization of Nucleic Acid

Andleeb Javaid, Nulevino Iralu, Sumiah Wani, Gowhar Ali, and Aflaq Hamid

Abstract

Polyacrylamide gel electrophoresis (PAGE) is a widely employed, versatile technique used in life sciences to fractionate and characterize molecules based on their size, shape, and charge. This high-resolution method primarily separates proteins in mixed samples, enabling their identification and quantification. While PAGE is predominantly used for protein analysis, it also separates nucleic acids, including DNA and RNA. This technique is particularly useful in resolving RNA-protein complexes and detecting RNA interactions through shifts in electrophoretic mobility. A key advantage of PAGE is its exceptional resolving power, capable of distinguishing between differing by as little as 1 in 500 base pairs.

Key words PAGE, Resolving power, Quantification

1 Introduction

The second most commonly used gel is polyacrylamide gel after agarose gel due to its high stability and ability to be created at gradient of concentrations or at wide range of concentrations. This is far more effective for the electrophoretic separation of nucleic acids and proteins. Polyacrylamide gels are most frequently used to examine proteins and chain-terminated DNA molecules because their resolving power is higher than that of agarose gels [1, 2]. When used for protein separations based on their molecular weight, it has high resolving power than sodium dodecyl sulfate-polyacrylamide gel electrophoresis (SDS-PAGE) [3]. Since the charge on proteins is more erratic, it is dependent on the pH, net charge, and content of amino acids [4]. Tethered silica nanoparticles were used in a recent procedure that involved the innovation of 3-[(methacryloxy)propyl] trimethoxysilane, which uses covalent bonds to lock the particles in the gel [5]. Acrylamide polymerization can result in much stronger gel that can be used for the

Aflaq Hamid et al. (eds.), *Detection of Plant Viruses: Advanced Techniques*, Springer Protocols Handbooks,
https://doi.org/10.1007/978-1-0716-4390-7_18,

electrophoretic separation of both proteins and nucleic acids. A small amount of acrylamide cross-linked by methylene bridge (N, N′ methylene bisacrylamide) facilitates the formation of a cross-linked gel that is both chemically and mechanically inert and has a highly regulated porosity. The polymerization is initiated by free radicals generated from ammonium persulfate (APS) and catalyzed by N,N, N′, N′-tetramethylethylenediamine (TEMED), which acts as an oxygen scavenger. The ratio of acrylamide: N, N'methylene bisacrylamide is generally 19:1 for DNA separation and 40:1 for protein separation [6]. The core principle of this PAGE technique is that, under an applied electrical potential, molecules are separated based on their mass-to-charge (m/z) ratio. The interaction of samples with the buffers, gel matrix, and electric field causes the particles to migrate at different rates according to their shape, charge, and size. This results in distinct bands representing individual molecules. A network of pores is created in the gel by cross-linking polyacrylamide and bisacrylamide and the size of these pores, which determines how molecules are sorted, is controlled by the concentration of the monomers (%T) [7]. Different types of polyacrylamide gel electrophoresis (PAGE) are used, including one-dimensional (1D) techniques like PAGE and isoelectric focusing (IEF), as well as two-dimensional (2D) PAGE.

In PAGE, proteins/DNA/RNA migrate through the gel matrix in response to an electric field, with the rate of migration determined by the gel pore size, as well as the DNA/RNA/protein size, shape, or charge. The pore size of gel decreases with increasing acrylamide concentration. Polyacrylamide gels, which run vertically and can vary in length, offer several advantages over agarose gels, including higher resolving power, the capacity to handle large amounts of DNA/RNA/proteins, and have ability to separate DNA/RNA/protein molecules that differ in length by as little as 0.2%. DNA/RNA/proteins collected from polyacrylamide gels are also highly pure and suitable for various applications.

2 Materials

2.1 Reagents

1. 1× TAE electrophoresis buffer.
2. Ammonium persulfate.
3. Tetramethylethylenediamine (TAMED) electrophoresis grade.
4. DNA loading buffer.
5. Deionized water.
6. Agarose stock solution for DNA gels.

2.2 Equipments

1. Laboratory wipes.
2. Beaker.
3. Gel plates.
4. Gloves.
5. Combs.
6. Ethanol, glass cleaner, soap.
7. Electrophoresis apparatus.
8. Pipette.
9. Microwave.
10. Imaging equipment.
11. Water bath.

3 Method

1. Clean the gel plates with ethanol to remove any dried acrylamide or particulate.
2. Make the desired gel solution using the above Table 1.
3. Stir the gel mixture.
4. Heat the mixture briefly to boil in water bath at 60 °C, but don't heat it for longer period because polymerization process requires the gel solution to remain at room temperature.
5. Gel solution should be filtered through three layers of Whatman No. 1 paper.
6. Set up the plates for gel mixture.
   (a) Put the spacer inside edge on long plate.
   (b) Place the second plate on top of the long plate with the agent side facing down.
   (c) Use binder clips to secure.

Table 1
Polyacrylamide gel solution

Gel %	Acrylamide (g)	Bisacrylamide (g)	Urea (g)	TAE 10× buffer (mL)	Deionized water (mL)	10% APS (μL)	TEMED (μL)
6	5.7	0.3	48	10	40	500	50
8	7.6	0.4	48	10	40	500	50
10	9.5	0.5	48	10	40	500	50
12	11.4	0.6	48	10	40	500	50
16	15.2	0.8	48	10	40	500	50

   (d) Tightly fit the comb.
   (e) Secure plates with tape along the edge.
7. Fill a 60 mL syringe with the gel solution after adding TEMED and ammonium persulfate.
8. Before the gel polymerizes, quickly fill the assembled plates.
   (a) Tip the plate so that the solution falls to the edge closest to you and then lay it horizontally.
   (b) Verify that the entire region is covered by the solution.
9. Make sure the absence of bubbles and firmly clamp the comb between the two plates.
10. Verify that the spaces between the wells created by the comb are appropriately formed after 30 min has passed.
11. Wait for 30 min more for complete polymerization.
12. Remove all clips and other items used to hold plates in place after the gel has solidified.
13. Rinse the top of gel after separating the plates.
14. Take the comb out and wash the comb region.
15. Fit the gel apparatus with gel plates.
16. Add TAE buffer.
17. Remove the lid and Flush wells with 1× TAE.
18. Add sample to the wells.
19. Allow to run for min at constant Watts (80 W) based on the size of samples.
20. Turn off the power supply.
21. Take the gel out of the spacers and apparatus.
22. Use spatula to carefully separate the gel from apparatus so that the gel is retained on the smaller plate.
23. The gel must now be exposed to the selected visualizing technique, such as CYBR Green or silver staining, as the final step.
    (a) Polysaccharide gels are formed by boiling followed by cooling. Interchain cross-linking is produced by rearranging of hydrogen bonds.
    (b) Agarose is composed of agarobiose.
    (c) Polyacrylamide gel is formed by polymerization of acrylamide. The polymerization reaction is initiated by persulfate radicals and catalyzed by TEMED.

Staining of Gel DNA molecules electrophoresed with the extrinsic fluoro-ethidium bromide can be easily seen under an ultraviolet lamp. Alternatively, nucleic acids can be stained by immersing the gel in ethidium bromide solution following electrophoretic separation. This molecule exhibits significantly increased fluorescence when intercalated into double-stranded DNA. The extrinsic fluor 1-anilino 8-naphthalene sulfonate can also be used to detect DNA.

4 Notes

1. Acrylamide concentration affects the resolution and separation range of proteins/DNA/RNA. Higher percentages resolve smaller molecules better, while lower percentages resolve larger ones [8].
2. Acrylamide is toxic before polymerization. Use gloves and work in well-ventilated area when handling [9].
3. Ammonium persulfate (APS) and TEMED are used to catalyze the polymerization of acrylamide. Freshly prepared APS is crucial for good polymerization [10].
4. Ensure proper mixing of APS and TEMED to avoid gel polymerization, which can affect migration patterns [11].
5. Common buffers include Tris-Glycine and Tris-Tricine. The choice of buffer influences resolution, e.g., Tris-Tricine improves separation of small proteins/peptides [12].
6. Ensure the pH of buffers is correct, as incorrect pH can affect stability and migration of proteins/DNA/RNA [13].
7. In SDS-PAGE, samples are treated with SDS and β-mercaptoethanol or DDT to ensure proteins are denatured and reduced. This linearizes proteins and adds a uniform negative charge [8].
8. Start with low voltage to avoid overheating and then increase once the samples have entered the gel. Higher voltages can lead to faster runs but may cause smearing if the gel overheats [14].
9. Ethidium bromide (EtBr) or SYBR Green is used for DNA/RNA visualization. Be cautious with EtBr, as it is mutagenic [10].
10. Ensure complete destaining of the gel when using stains like Coomassie to improve the contrast and visibility of bands [15].
11. Polyacrylamide gels are fragile, especially at lower acrylamide concentrations. Handle the gels carefully to avoid tearing during staining or transfer [16].
12. Uneven band migration can happen if the gel is not evenly polymerized, the buffer is old or the gel is running at too high voltage [11].

References

1. Anderson J, Wright D, Meksem K (2013) Agarose gel electrophoresis and polyacrylamide gel electrophoresis for visualization of simple sequence repeats. Methods Mol Biol 1006:167
2. Romano EF, Quirino JP (2018) Frontal analysis capillary electrophoresis: recent advances and future perspectives. Bioanalysis 10(14): 1143–1159
3. Brody JR, Kern SE (2004) History and principles of conductive media for standard DNA electrophoresis. Anal Biochem 333(1):1–13
4. Lodish H, Berk A, Matsudaira P, Kaiser CA, Krieger M, Scott MP, Darnell J (2004) Molecular cell biology, 4th edn
5. Hanauer M, Pierrat S, Zins I, Lotz A, Sönnichsen C (2007) Separation of nanoparticles by gel electrophoresis according to size and shape. Nano Lett 7(9):2881–2885
6. Barasinski M, Garnweitner G (2020) Restricted and unrestricted migration mechanisms of silica nanoparticles in agarose gels and their utilization for the separation of binary mixtures. J Phys Chem C 124(9):5157–5166
7. Packirisamy V, Pandurangan P (2023) Polyacrylamide gel electrophoresis: a versatile tool for the separation of nanoclusters. BioTechniques 74(1):51–62
8. Laemmli UK (1970) Cleavage of structural proteins during the assembly of the head of bacteriophage T4. Nature 227(5259):680-685
9. OSHA (2020) Acrylamide. Occupational safety and health administration (OSHA)
10. Sambrook J, Russell DW (2001) Molecular cloning: a laboratory manual (3rd eds). Cold spring harbor laboratory press
11. Current Protocols in Molecular Biology (2001) Gel electrophoresis of proteins. Wiley
12. Farrell PH (1975) High resolution two-dimensional electrophoresis of proteins. J Biol Chem 250(10):4007-4021
13. Bradford MM (1976) A rapid and sensitive method for the quantitation of microgram quantities of protein utilizing the principle of protein-dye binding. Anal Biochem 72(1-2): 248-254
14. Neuhoff V, Stamm R, Eibl, H (1988) The electrophoresis of proteins in polyacrylamide gels. Electrophoresis 9(2):93-99
15. Schägger H, von Jagow G (1987) Tricine-SDS-PAGE: a novel method for resolving proteins in polyacrylamide gels at low polyacrylamide concentrations. Electrophoresis 8(2): 122-128
16. Harlow E Lane D (1988) Antibodies: a laboratory manual. Cold spring harbor laboratory press

Chapter 19

Measurement of RNA/DNA Concentration Using Spectrophotometer

Insha Mehraj, Nulevino Iralu, Sumiah Wani, Sahar Saleem, and Aflaq Hamid

Abstract

Precise measurement of nucleic acid concentration is crucial for various applications such as transinfection, cloning, PCR, and next-generation sequencing (NGS). Achieving the specified target concentration is essential for optimal performance in these processes. Inaccurate quantification may introduce variability in downstream assays, impacting result quality. DNA and RNA quantification can be conducted through methods like UV absorbance through a spectrophotometer like Nanodrop, fluorescence measurement with nucleic acid-binding dyes, and qPCR. The appropriate DNA or RNA quantification method should be chosen based on many factors including availability of equipment, intended downstream application, and throughput. Some quantification methods have the added benefit of enabling determination of the biological and chemical purity of a sample. Caution should be used when comparing yields between methods as the presence of contaminants can lead to interference in measurements, resulting in over- or under-estimation of yield.

Key words RNA/DNA quantification, Spectrophotometer, Nanodrop

1 Introduction

The primary method employed to assess the concentration and purity of nucleic acids is absorbance. Absorbance measurements are utilized for estimating the concentration of DNA or RNA in purified samples. In this approach, the nucleic acid sample is inserted into a quartz cuvette, which is subsequently positioned within a UV spectrophotometer [1]. Compact spectrophotometers like the NanoDrop™ device, DeNovix DS-11 spectrophotometer, and Blue-Ray Bio EzDrop spectrophotometer facilitate the examination of sample volumes as minimal as 1 μL using pedestals. UV light is transmitted through the sample at a specified path length, and concentrations of nucleic acids are directly computed by

Aflaq Hamid et al. (eds.), *Detection of Plant Viruses: Advanced Techniques*, Springer Protocols Handbooks,
https://doi.org/10.1007/978-1-0716-4390-7_19,

gauging absorbance values at 260 nm against a blank using Beer-Lambert's equation [2]:

$$A = \varepsilon c l$$

where A = UV absorbance

ε = wavelength-dependent extinction coefficient,

c = nucleic acid concentration,

l = light path (cm).

2 Extinction Coefficients*

dsDNA (pure): 0.020 $(\mu g/mL)^{-1}\ cm^{-1}$,
ssDNA (pure): 0.027 $(\mu g/mL)^{-1}\ cm^{-1}$,
ssRNA (pure): 0.025 $(\mu g/mL)^{-1}\ cm^{-1}$ [3].

*It's crucial to note that this formula is applicable solely to substantial nucleic acid molecules with proportional nucleotide compositions, such as genomic DNA or plasmids. For oligonucleotides and other short nucleic acid molecules like miRNAs, the extinction coefficient should be determined based on the oligonucleotide sequence.

Quantification of RNA/DNA can be carried out by checking its absorbance at wavelength of 260 nM or 280 nM. The absorbance peak of nucleic acid is 260 nM. If the sample is reasonably pure, its concentration can be measured using UV absorbance spectrophotometry because the base pairs in nucleic acid can absorb UV light, and hence the amount of base pairs is directly proportional to DNA concentration. For double-stranded DNA, sample containing pure dsDNA will give an absorbance of 1 at 260 nM, which is approximately 50 ug/mL of dsDNA. For single-stranded DNA, sample will show 1 at 260 nM, which is approximately 33 ug/mL of DNA. For pure RNA, sample will show approximately 40 ug/mL of RNA [4–7].

For enhanced precision, the A260 measurement is frequently adjusted for turbidity (measured by absorbance at 320 nM) using the subsequent equation:

$$\begin{aligned}&\text{Concentration } (\mu g/mL)\\&\quad = (\text{A260 measurement–A320 measurement})\\&\quad \times \text{nucleic acid conversion factor}^{**} \times \text{dilution factor}\end{aligned}$$

**Conversion factor for dsDNA: 50 μg/mL

**Conversion factor for ssDNA: 37 μg/mL

**Conversion factor for ssRNA: 40 μg/mL.

Total yield is then derived by multiplying the nucleic acid concentration by the final total purified sample volume.

$$\text{DNA yield } (\mu g) = \text{DNA concentration} \times \text{total sample volume (mL)}$$

Light absorption within the 260 nM range can be influenced by molecules other than DNA or RNA. Aromatic ring-containing amino acids in proteins, such as those present in guanidine and other chaotropic salts or solvents, commonly used in silica-based DNA and RNA purification, absorb light at 280 nM and 230 nM, respectively. This absorption interference can result in higher 260 nM absorbance readings due to crossover effects. Consequently, the presence of contaminants can lead to an overestimation of DNA quantity if the A260 value is utilized in yield calculations.

To assess protein contamination, the ratio of absorbance at 260 nM (representing nucleic acids) to 280 nM (representing aromatic ring absorption in protein amino acids, with phenol also absorbing at 280 nM) is determined. High-quality DNA typically exhibits an A260/A280 ratio of 1.7–2.0, while high-quality RNA shows an A260/A280 ratio of approximately 2.0 [8].

$$\text{DNA purity (protein contaminants)} = \text{A260 reading} \div \text{A280 reading}$$

For the assessment of chemical contamination, the ratio of absorbance at 260–230 nm is employed. Chaotropic salts and organic solvents, known to absorb light at 230 nm and potentially inhibit PCR, can leave residues in DNA or RNA preparations. In applications sensitive to chemical interference, such as enzyme-based assays like qPCR and sequencing, a desirable A260/A230 ratio should be greater than 1.5. A lower ratio indicates a higher presence of organic compounds or chaotropic salts in the sample.

$$\text{DNA purity (chemical contaminants)} = \text{A260 reading} \div \text{A230 reading}$$

3 Materials

1. DNA/RNA sample (*see* **Note 1**).
2. Nuclease-free water.
3. UV spectrometer (NanoDrop).

4 Methods

1. Dilute 5 uL of sample with 995 uL of nuclease-free water (*see* **Note 2**).

2. Calibrate spectrophotometer at 260 nM and 280 nM to zero using water blanks (*see* **Note 3**).
3. Pour the diluted sample into the cuvette and note the measurements at 260 nM and 280 nM (*see* **Note 4**).
4. Take multiple readings to avoid errors and take the average of both.
5. Calculate the DNA concentration using given formula.

4.1 Calculation

DNA (ug/mL)	$\frac{(OD_{260}) \times (\text{Dilution factor}) \times (50\ \text{ug DNA/mL})}{(1\ OD_{260}\text{unit})}$
RNA (ug/mL)	$\frac{(OD_{260}) \times (\text{Dilution factor}) \times (40\ \text{ug DNA/mL})}{(1\ OD_{260}\text{unit})}$

4.2 To Check Purity

UV spectrophotometer can be used to check the purity of a nucleic acid solution by measuring the absorbance of solution at two wavelengths, i.e., 260 nM and 280 nM, and calculating their ratio (*see* **Note 5**).

$$\text{Purity} = A_{260}/A280.$$

For DNA, ratio value of A_{260}/A_{280} at **1.8** is a characteristic of pure DNA.

For RNA, ratio value of A_{260}/A_{280} at **2** is a characteristic of pure RNA.

5 Notes

1. Ensure that the RNA/DNA sample is free from contaminants like proteins, phenols, or salts. These can interfere with absorbance readings.
2. Dilute the sample appropriately to bring it within the spectrophotometer's linear detection range. Use nuclease-free water or a compatible buffer.
3. Regularly calibrate the spectrophotometer using a blank. Always blank the instrument with the same solution used to prepare the sample.
4. Ensure the cuvette or Nano-Drop pedestal is clean and free from residues. Contaminants can alter readings.
5. To assess protein contamination, the ratio of absorbance at 260 nM (representing nucleic acids) to 280 nM (representing aromatic ring absorption in protein amino acids, with phenol also absorbing at 280 nM) is determined.

References

1. Deprez L, Corbisier P, Kortekaas AM, Mazoua S, Beaz Hidalgo R, Trapmann S, Emons H (2016) Validation of digital PCR method for quantification of DNA copy number concentrations by using a certified reference material. Biomol Detect Quantif 9:29–39
2. Rafferty JP (2024) Beer's law. Encyclopedia Britannica
3. Anonymous (2013) A guide to understandanding extinction coefficients, with emphasis on spectrophotometric determination of protein concentration. Piecrce Biotechnology (ThermoScientific)
4. Desjardins R, Conklin DS (2011) Microvolume quantition of nucleic acid. Curr Protoc Hum Genet 93(1)
5. Ellison SLR, Williams A (2012). Quantifying uncertainty in analytical measurement. Eurachem/Citac Guide CG, vol. 1, issue 3, p 141
6. Forootan A, Sjoback R, Bjorkman J, Sjogreen B, Linz L, Kubista M (2017) Methods to determine limit of detection and limit of quantification in quantitative real-time PCR (qPCR). Biomol Detect Quantif 12:1–6
7. Garcia-Alegria, A.M., Anduro-Corona, I., Perez-Martinez, C.J., Corella-Maduena, M.A., Rascon-Duran, M.L. and Astiazran-Gracia, H. (2020). Quantification of DNA through the NanoDrop Spectrophotometer:methodological validation using standard reference material and Sprague Dawley rat and human DNA. Int J Anal Chem
8. Gill SC, Hippel PH (1989) Calculation of protein extinction coefficients from amino acid sequence data. Anal Biochem 182:319–326

Chapter 20

Polymerase Chain Reaction (PCR) for Virus Detection

Nulevino Iralu, Sumiah Wani, Dasari Meghanath, and Aflaq Hamid

Abstract

Polymerase chain reaction (PCR) is essential in plant virology for its ability to amplify and detect specific DNA or RNA sequences with high sensitivity and specificity. It enables early and accurate detection of plant viruses by amplifying even minute amounts of viral genetic material, which is crucial for managing and preventing the spread of infections. PCR's specificity allows for precise identification of different viral strains and genomes, while quantitative PCR (qPCR) measures viral load, providing insights into infection severity. This technique is integral to diagnosing viral infections reliably and quickly, which is vital for effective disease management.

Key words PCR, Amplification, Primers, Template, Temperature

1 Introduction

PCR supports research into viral genetics and virus–host interactions, contributing to the development of resistant plant varieties and innovative control strategies [1]. Its capability to screen large numbers of samples efficiently is also crucial for maintaining virus-free plant stock in breeding and production programs. Overall, PCR is a fundamental tool in plant virology, driving advances in both research and practical disease management [2]. This reaction is accomplished in three steps: denaturation (above 95 °C), annealing (50–70 °C), and extension (72 °C), which functions in cycles using a DNA template, primers, and DNA polymerase to create multiple copies of a specific DNA segment. Annealing temperature is decided based on the primers used [3].

2 Materials

1. Forward primer.
2. Reverse primer.

Aflaq Hamid et al. (eds.), *Detection of Plant Viruses: Advanced Techniques*, Springer Protocols Handbooks,
https://doi.org/10.1007/978-1-0716-4390-7_20,

3. DNA sample.
4. GoTaq green.
5. Nuclease-free water.

3 Procedure (*see* Notes 1–4)

3.1 Master Mix Preparation (see Notes 5 and 6)

1. The quantity of master mix will depend upon the number of samples to be processed (*see* **Note 9**).

Standard protocol 25 uL for 1× reaction	
GoTaq green	12.5 uL
Reverse primer	1 uL
Forward primer	1 uL
Nuclease-free water	8.5 uL
Total	23 uL

2. Take a microtube and add GoTaq Green, forward primer, reverse primer, and nuclease-free water. Keep the master mix in ice.
3. Take 2 uL of DNA template in PCR tube and add 1× reaction mixture to each tube from the master mix tube. Keep in ice to avoid degradation.

3.2 PCR Program (see Notes 7 and 8)

S. No	Stage	Temperature (°C)	Time
1.	Initial denaturation	95	5 min
2.	Denaturation	95	1 min
3.	Annealing	50–60 (based on primers)	1.5 min
4.	Extension	72	1 min
5.	Final extension	72	7 min

1. Stage 2–4 should be repeated for 30 cycles.
2. After setting the PCR program, start the program. Once the program is over, keep the PCR tubes in ice and load in 1% agarose gel, then visualize under gel doc (*see* **Notes 10** and **11**).

4 Notes

1. Maintain a clean working environment. Contamination is a significant risk in PCR, as even minute amounts of DNA from previous experiments can lead to false positives.
2. Use dedicated PCR equipment and consumables, such as pipettes, tubes, and tips, that are reserved exclusively for PCR work.
3. Regularly clean surfaces with suitable disinfectants and use disposable gloves to further minimize contamination risks.
4. Meticulous preparation of reagents: all components, including primers, nucleotides, and enzymes, should be freshly prepared or properly stored to prevent degradation.
5. Use high-quality, enzyme-grade reagents and avoid reusing aliquots of PCR mixes or master mixes, as this can introduce contamination.
6. Setting up PCRs in a designated area separate from post-PCR analysis areas to prevent the introduction of amplified products into the reaction setup.
7. Maintain precise temperature and timing during the PCR process is vital for successful amplification.
8. Using a well-calibrated thermocycler and regularly verifying its performance can help ensure that denaturation, annealing, and extension steps occur at the correct temperatures.
9. Following the manufacturer's protocols for the specific polymerase being used, including optimal extension times and temperatures, is also essential for achieving consistent results.
10. Proper handling and disposal of PCR products are critical. Once the PCR is complete, products should be analyzed in a designated area equipped with appropriate safety measures, such as UV light to help degrade any residual DNA.
11. All materials that have come into contact with PCR products should be disposed of according to biohazard regulations to prevent environmental contamination and safeguard laboratory personnel.

References

1. Gaur RK, Khurana SP, Sharma P, Hohn T (eds) (2021) Plant virus-host interaction: molecular approaches and viral evolution. Academic Press
2. Rubio L, Galipienso L, Ferriol I (2020) Detection of plant viruses and disease management: relevance of genetic diversity and evolution. Front Plant Sci 11:1092
3. Lorenz TC (2012) Polymerase chain reaction: basic protocol plus troubleshooting and optimization strategies. J Vis Exp 63:e3998

Chapter 21

Reverse Transcription Polymerase Chain Reaction (RT-PCR) for RNA Viruses

Nulevino Iralu, Dasari Meghanath, Sumiah Wani, Shahjahan Rashid, Wani Farhana, and Aflaq Hamid

Abstract

The prompt and precise identification of plant viruses is crucial for effective disease management and crop protection. Reverse transcription polymerase chain reaction (RT-PCR) has become a vital tool in plant virology, allowing for the specific detection of viral RNA in infected plant tissues. This method amplifies viral genetic material, offering a level of sensitivity and specificity that outperforms traditional diagnostic techniques. RT-PCR enables early detection of viral pathogens, which is essential for timely intervention and minimizing crop damage. Through its continued evolution, RT-PCR remains a fundamental component in combating plant viral diseases, supporting sustainable agriculture and food security.

Key words PCR, RNA, Virus, Reverse transcription, Primers

1 Introduction

Reverse transcription polymerase chain reaction (RT-PCR) is a molecular biology technique that has revolutionized the field of plant virology by providing a reliable and efficient method for detecting viral infections. RT-PCR enables the direct amplification of viral RNA, allowing for precise identification of a wide range of viral pathogens at early stages of infection [1]. In recent years, the increasing prevalence of plant viruses, which can cause significant crop losses and threaten food security, has highlighted the urgent need for advanced diagnostic tools. RT-PCR not only enhances sensitivity and specificity in virus detection but also allows for the differentiation between closely related viral species and strains [2, 3]. This capability is crucial for effective disease management and the development of resistant plant varieties. Because of its versatility, RT-PCR has become an indispensable tool in virus monitoring, epidemiological studies, and breeding programs [4].

Aflaq Hamid et al. (eds.), *Detection of Plant Viruses: Advanced Techniques*, Springer Protocols Handbooks,
https://doi.org/10.1007/978-1-0716-4390-7_21,

2 Materials

1. RNA Extraction Kits: Kits designed for isolating high-quality total RNA from plant tissues.
2. Reverse Transcriptase Enzyme: Enzyme required for synthesizing complementary DNA (cDNA) from RNA.
3. PCR Master Mix: Contains DNA polymerase, dNTPs (deoxynucleotide triphosphates), buffer, and magnesium ions.
4. Primers: Specific forward and reverse primers designed for the target viral RNA sequence.
5. Nuclease-Free Water: To dilute reagents and for reaction setup, ensuring no contamination.
6. Thermal Cycler: Equipment for performing the temperature cycling necessary for both reverse transcription and PCR amplification.
7. RNAse Inhibitors (optional): To protect RNA from degradation during the process.
8. Gel Electrophoresis Equipment (for analysis).
9. DNA Ladder (size marker): For determining the size of amplified products during gel analysis.
10. Loading Dye: To visualize samples during gel electrophoresis.
11. UV Transilluminator or Gel Imaging System.

3 Method

3.1 RNA Extraction

Using the standard protocol of the RNA extraction kit, RNA extraction can be performed (refer Chap. 16) (*see* **Notes 1–6**).

3.2 cDNA Synthesis

1. cDNA was generated in a reverse transcription reaction using total nucleic acid and a reverse primer. This mixture was incubated at 95 °C for 10 min, followed by rapid cooling on ice for 1 min or storage in a deep freezer at −80 °C.
2. A master mix was prepared, consisting of 5x RT strand buffer, dNTP mix, ribonuclease inhibitor, and MMLV RT-polymerase. This mixture was evenly distributed into each tube containing nucleic acid and gently mixed.
3. Tubes containing the reaction mixture was incubated for 1 h at 37 °C followed by 15 min at 72 °C in a thermocycler.

Component	Volume (l)
5× First strand buffer	*
10 mM dNTP	*

(continued)

Component	Volume (l)
20 U Ribonuclease inhibitor	*
200 U M-MLV reverse transcriptase	*
10 mM reverse primer	1 μL
Template RNA	5 μL

* as per the manufacturer instruction

3.3 RT-PCR

(a) Mastermix preparation: The quantity of master mix will depend upon the number of samples to be processed. Reaction mixture using the synthesized cDNA was made:

Component	Volume
cDNA	2 μL
10 mM forward primer	2 μL
10 mM reverse primer	2 μL
10× Taq buffer	2.5 μL
25 mM mgcl2	0.5 μL
10 mMdNTP	0.5 μL
2.5 U Taq DNA polymerase	0.5 μL
Total	10 μL + 15 μL nuclease free water

(b) Keep in ice to avoid degradation.

(c) PCR program: The temperature and duration of the PCR program (especially annealing) will entirely depend on primer being used. Table 1 below is for reference (*see* **Note 7**).

(d) After setting the PCR program, start the program.

Table 1
Primer, temperature, and duration

S. No	Stage	Temperature (°C)	Time
1.	Initial denaturation	95	5 min
2.	Denaturation	95	1 min
3.	Annealing	50–60 (based on primers)	1.5 min
4.	Extension	72	1 min
5.	Final extension	72	7 min

3.4 Visualization

Once the program is over, keep the PCR tubes in ice and load in 1% agarose gel, then visualize under gel doc (refer Chap. 17) (*see* **Notes 8–9**).

4 Notes

1. RNA handling: Use RNase-free equipment and reagents to prevent degradation of RNA. Wear gloves and change them frequently to avoid contamination.
2. Work in a clean environment: Perform all steps in a designated RNA work area or a laminar flow hood. Regularly clean surfaces with RNase decontamination solutions.
3. Use nuclease-free reagents: Ensure that all reagents, including water and buffers, are nuclease-free.
4. Prepare separate work areas: Set up distinct areas for RNA extraction, cDNA synthesis, and PCR amplification to minimize cross-contamination.
5. Prevent contamination of samples: Use aerosol-resistant pipette tips and dedicated pipettes for RNA work. Always change tips between samples.
6. Include controls: Incorporate positive and negative controls in each run to validate the results and detect contamination.
7. Optimize reaction conditions: Carefully optimize temperatures and times for reverse transcription and PCR to ensure efficient amplification.
8. Storage conditions: Store RNA samples at −80 °C and minimize freeze-thaw cycles to preserve integrity.
9. Documentation: Keep thorough records of all reagents, samples, and experimental conditions to trace any potential issues.

References

1. Yadav N, Khurana SP (2016) Plant virus detection and diagnosis: progress and challenges. In: Frontier discoveries and innovations in interdisciplinary microbiology, pp 97–132
2. Arous S, Harmon CL, Capobianco HM, Polston JE (2018) Comparison of genus-specific primers in RT-PCR for the broad-spectrum detection of viruses in the genus Potyvirus by plant diagnostic laboratories. J Virol Methods 258:29–34
3. Cassedy A, Parle-McDermott A, O'Kennedy R (2021) Virus detection: a review of the current and emerging molecular and immunological methods. Front Mol Biosci 8
4. Wani S, Saleem S, Nabi SU, Ali G, Paddar BA, Hamid A (2021) Distribution and molecular characterization of potato virus Y (PVY) strains infecting potato (Solanum tuberosum) crop in Kashmir (India). VirusDisease 32:78–788, 637559

Chapter 22

RT-Nested PCR for Plant Virus Detection

Dasari Meghanath, Sumiah Wani, Nulevino Iralu, and Aflaq Hamid

Abstract

RNA viruses are a diverse group of viruses responsible for a wide range of plant diseases and a major threat to plant health. The procedure for a simple, ultrasensitive, and specific method using a reverse transcriptase-nested polymerase chain reaction is provided for the detection of RNA viruses from naturally infected plants. This protocol consists of RNA extraction, cDNA synthesis, RT-PCR, and RT-nPCR. Two sets of primer pairs were used in which the outer primers amplify a large fragment of target DNA and nested primers amplify a smaller, more specific region within the first amplicon. This is a powerful tool for amplifying RNA targets with high sensitivity and specificity and could be a promising diagnostic application.

Key words RNA viruses, RNA isolation, cDNA synthesis, RT-PCR, RT-nPCR

1 Introduction

ELISA, immunofluorescence, RT-PCR, m-PCR, and cDNA probes [2, 3, 7, 10, 12, 13] are various methods used to detect plant viruses. ELISA is the most widely used serological method for a single virus detection [9] from past to present. Currently, due to the availability of viral genetic information, molecular methods are most commonly used for diagnosis of viruses due to its higher accuracy and more specificity. Different variants of PCR like Multiplex PCR, qPCR, RT-PCR are used to identify different plant viruses. An already existing variant of PCR, called the nested PCR, enhances sensitivity and specificity by using two sets of primers, one nested with the other. This method is useful for detecting low-abundance targets, unstable viral RNA, mutated virus, and those that cannot be visualized due to low amplification product [13]. These above problems can be countered by nested PCR [11]. The product from the first round PCR amplification is used as the template for a second PCR amplification. This second reaction uses a second set of primers (nested primers) that bind within the amplicon produced in the first round. However, the second

Aflaq Hamid et al. (eds.), *Detection of Plant Viruses: Advanced Techniques*, Springer Protocols Handbooks,
https://doi.org/10.1007/978-1-0716-4390-7_22,

reaction faces the increased time, cost, and risk of contamination [8]. Several low titer viruses such as PDV, PNRSV, PPV, and CTV were detected by this technique [1, 4, 5, 6, 11]. This nested PCR combination with Immunocapture-RT PCR increases sensitivity and simplify sample preparation [4, 5]. In this experiment, nested RT-PCR protocol was described to detect RNA viruses, but the same procedure can be done in a conventional nested PCR to detect DNA viruses without the need for reverse transcription.

2 Materials

1. Overnight sterilization of glassware, mortar, and pestles at 180 °C.
2. Sterilized eppendorf tubes, pipettes, and tips.
3. 1 mL of diethyl pyrocarbonate (DEPC) is added to 1 L of double distilled water, incubated overnight, and autoclaved at 120 °C for 20 min.
4. Extraction buffer: 100 mM Tris–HCl (pH 8.0), 2% w/v cetyltrimethylammonium bromide (CTAB), 25 mM EDTA (pH 8.0), 2.5 M NaCl, 2% w/v PVP were autoclaved at 120 °C for 20 min.
5. SSTE buffer: 10 mM Tris–HCl (pH 8.0), 1 mM EDTA (pH 8.0), 1% SDS, 1 M NaCl. Do not autoclave.
6. β-Mercaptoethanol, chloroform–isoamyl alcohol (24:1 v/v) and isopropanol (at −20 °C).
7. LiCl dissolved in DEPC water at 10 M and autoclaved at 120 °C for 20 min (store at 4 °C).
8. Sodium acetate is dissolved in DEPC water at 3 M concentration; Adjust the P^{H} to 5.8 by adding acetic acid. Autoclave at 120 °C for 20 min.
9. 70% Ethanol prepared with DEPC water. Store at −20 °C.

2.1 cDNA Synthesis

1. Template RNA.
2. 10 μM reverse primer.
3. dNTP mix: 10 mM of each dNTP.
4. 100/200 U MMLV reverse transcriptase.
5. 5× RT first strand buffer.
6. 20 U Ribonuclease inhibitor.

2.2 RT-PCR Amplification

1. cDNA.
2. dNTP mix: 10 mM of each dNTP.
3. 10× Taq buffer.

4. 25 mM Mgcl$_2$.
5. 2.5/5 U Taq DNA polymerase.
6. Forward and reverse external primers (5 μM).

2.3 One-Step RT-PCR

1. Template RNA.
2. dNTP mix: 10 mM of each dNTP.
3. 10× Taq buffer.
4. 25 mM Mgcl$_2$.
5. 2.5/5 U Taq DNA polymerase.
6. Forward and reverse external primers (5 μM).
7. 100/200 U MMLV reverse transcriptase.

2.4 Nested RT-PCR

1. First round PCR product.
2. dNTP mix: 10 mM of each dNTP.
3. 10× Taq buffer.
4. 25 mM Mgcl$_2$.
5. 2.5/5 U Taq DNA polymerase.
6. Forward and reverse internal primers (5 μM).

3 Methods

3.1 RNA Extraction

1. Prepare a pre-warmed extraction buffer (65 °C) with freshly added β- mercaptoethanol (20 μL/mL).
2. Grind 0.2–0.5 g of leaf tissue in liquid nitrogen using a sterile mortar and pestle in 2 mL micro-centrifuge tubes.
3. Using a sterile spatula, transfer the ground tissue powder into a 2 mL Eppendorf tube containing 800 μL of extraction buffer with β-mercaptoethanol.
4. Vortex and incubate at 65 °C for 10 min, with intermittent vortexing again for every 3 min.
5. Add equal volume of chloroform–isoamyl alcohol (24:1), vortex vigorously, and centrifuge at 10,000 × g for 10 min at 4 °C.
6. Carefully pipette the upper aqueous phase to a new, clean microcentrifuge tube. Add one volume of chloroform–isoamyl alcohol (24:1) mixture to the aqueous phase, vortex vigorously, and centrifuge at 10,000 × g for 10 min at 4 °C.
7. Transfer aqueous phase to new 1.5 ml tube. Add 0.5 vol. of LiCl 10 M. Mix (invert) and precipitate on ice for 30 min.
8. Centrifuge for 30 min at 4 °C (20,000 × g).
9. Discard supernatant, vortex pellet with 500 μL pre-warmed SSTE (65 °C).

10. Add equal volume of chloroform–isoamyl alcohol (24:1) to the micro-centrifuge tube, vortex for 10 s and centrifuge at 10,000 × g for 10 min at 4 °C.
11. Recover the aqueous phase and add 0.5 V of isopropanol, 0.1 V of 3 M sodium acetate. Gently mix the tube by inverting.
12. Centrifuge at 20,000 × g for 20 min at 4 °C.
13. Wash the pellets with 70% ethanol and centrifuge at 21,000 × g for 5 min at 4 °C.
14. Remove the ethanol using a pipet after repeated pulse centrifugations and briefly air-dry the pellet.
15. Dissolve the pellet in 20 μL of DEPC-treated water and store at −80 °C.
16. Quantify and assess RNA purity with NanoDrop or by using spectrophotometer.

3.2 cDNA Synthesis

1. cDNA was synthesized in a reverse transcription reaction using total nucleic acid and a reverse primer. The reaction was incubated at 95 °C for 10 minutes, followed by immediate quenching on ice or in a deep freezer at −80 °C.
2. Master mixture was prepared containing 5× RT strand buffer, dNTP mix, ribonuclease inhibitor, MMLV RT-polymerase, and equally distributed to each tube containing nucleic acid and mixed gently.
3. Tubes containing the reaction mixture was incubated for 1 h at 37 °C followed by 15 min at 72 °C in a thermocycler.

Component	Volume (l)
5× First strand buffer	*
10 mM dNTP	*
20 U Ribonuclease inhibitor	*
200 U M-MLV reverse transcriptase	*
10 mM reverse primer	1 μL
Template RNA	5 μL

* as per the manufacturer instruction

3.3 First Round RT-PCR

First round PCR was performed by using cDNA in a reaction mixture containing Taq DNA polymerase, 10× PCR buffer, $Mgcl_2$, dNTP mix, and external forward and reverse primers.

Component	Volume
cDNA	2 μL
10 mM forward primer	2 μL
10 mM reverse primer	2 μL
10× Taq buffer	2.5 μL
25 mM Mgcl$_2$	0.5 μL
10 mM dNTP	0.5 μL
2.5 U Taq DNA polymerase	0.5 μL
Total	10 μL + 15 μL nuclease-free water

3.4 Nested RT-PCR

1. The initial RT-PCR product, approximately 2 μL, can be used directly as a template for nested PCR without any dilution.
2. Alternatively, the product can be diluted to various concentrations, such as 2:10, 2:20, and 2:30, to find the optimal dilution for efficient amplification in the nested PCR.
3. The dilution that yields the best results in the nested PCR will be considered the standard dilution point for subsequent reactions.
4. Nested PCR was performed by using PCR product of the first round, in a reaction mixture containing Taq DNA polymerase, 10× PCR buffer, Mgcl$_2$, dNTP mix, and nested(internal) forward and reverse primers.

Component	Volume
DNA	2 μL
10 mM Forward primer	2 μL
10 mM Reverse primer	2 μL
10× Taq buffer	2.5 μL
25 mM Mgcl$_2$	0.5 μL
10 mM dNTP	0.5 μL
2.5 U Taq DNA polymerase	0.5 μL
Total	10 μL + 15 μL nuclease-free water

Note: In both direct and nested PCR, use positive and negative controls to ensure PCR accuracy and to troubleshoot any potential issues

4 Notes

1. Minimize RNA contamination by working in RNase-Free Environment and by wearing gloves and lab coat and consider wearing a face mask to minimize contamination of your skin and breath.
2. Work in a Clean Zone designated as RNA-free workspace or laminar flow hood to minimize airborne RNases.
3. Use disposable, RNase-free pipette tips, tubes, and containers for all reagents and samples. Treat reusable equipment with RNase decontamination solutions.
4. Use DEPC-treated water or commercially available RNase-free water for all solutions to inactivate any lingering RNases.
5. Work quickly throughout the isolation process. If a delay is unavoidable, flash-freeze the tissue in liquid nitrogen or store it in RNA later solution to prevent RNA degradation.
6. Maintain a cold and sterile environment throughout. The addition of glass beads will facilitate aseptic grinding of leaf sample in a tissue lyser.
7. To ensure accurate and reliable PCR results, prioritize these steps to minimize contamination risks throughout reaction setup.
   - Use only sterilized and accurate pipettes.
   - Use only sterilized tips and a new tip for every pipetting step.
   - Following removal of the aliquot, ensure that the reagent/ sample tube is properly capped.
   - Use sterile and clean plasticware.
8. The following steps should be performed in a sterile environment:
   - Reaction cocktail (Mix) preparation with no RNA.
   - Addition of RNA (Laminar Air-flow chamber).

References

1. Adkar-Purushothama CJ, Maheshwar PK, Sano T, Janardhana GR (2011) A sensitive and reliable RT-nested PCR assay for detection of citrus tristeza virus from naturally infected citrus plants. Curr Microbiol 62:1455–1459
2. Brlansky RH, Lee RF, Garnsey SM (1988) In situ immunofluorescence for the detection of citrus tristeza virus inclusion bodies. Plant Dis 72:1039–1041
3. Garnsey SM, Cambra M (1991) Enzyme-linked immunosorbent assay (ELISA) for citrus pathogens. In: Roistacher CN (ed) Graft transmissible diseases of citrus. Handbook for detection and diagnosis. FAO, Rome, pp 193–216
4. Helguera PR, Docampo DM, Nome SF, Ducasse DA (2002) Enhanced detection of Prune dwarf virus in peach leaves by immunocapture-reverse

transcription-polymerase chain reaction with nested polymerase chain reaction (IC-RT-PCR nested PCR). J Phytopathol 150:94–96
5. Helguera PR, Taborda R, Docampo DM, Ducasse DA (2001) Immunocapture reverse transcription-polymerase chain reaction combined with nested PCR greatly increases the detection of Prunus necrotic ring spot virus in the peach. J Virol Methods 95:93–100
6. Huang Z, Rundell PA, Guan X (2004) Detection and isolate differentiation of citrus tristeza virus in infected field trees based on reverse transcription polymerase chain reaction. Plant Dis 88:625–629
7. Lopez MM, Llop P, Olmos A, Marco-Noales E, Cambra M, Bertolini E (2008) Are molecular tools solving the challenges posed by detection of plant pathogenic bacteria and viruses. Curr Issues Mol Biol 11:13–45
8. McLaughlin MR, Barnett OW, Burrows PM, Baum RH (1981) Improved ELISA conditions for detection of plant viruses. J Virol Methods 3:13–25
9. Narvaez G, Skander BS, Ayllon MA (2000) A new procedure to differentiate citrus tristeza virus isolates by hybridization with digoxigenin-labelled cDNA probes. J Virol Methods 85:83–92
10. Olmos A, Bertolini E, Cambra M (2002) Simultaneous and co-operational amplification (co-PCR): a new concept for detection of plant viruses. J Virol Methods 106:51–59
11. Olmos A, Cambra M, Esteban O, Gorris MT, Terrada E (1999) New device and method for capture, reverse transcription, and nested PCR in a single closed-tube. Nucleic Acid Res 27: 1564–1565
12. Roy A, Fayad A, Barthe G (2005) A multiplex polymerase chain reaction method for reliable, sensitive and simultaneous detection of multiple viruses in citrus trees. J Virol Methods 129: 47–55
13. Webster CG, Wylie JS, Jones MGK (2004) Diagnosis of plant viral pathogens. Curr Sci 86:1604–1607

Chapter 23

Real-Time PCR (qPCR) for Plant Virus Detection and Expression Analysis

Wani Farhana, Nulevino Iralu, Sumiah Wani, Dasari Meghanath, and Aflaq Hamid

Abstract

Real-time polymerase chain reaction (PCR), or quantitative PCR (qPCR), represents a ground-breaking advancement over traditional PCR techniques. By incorporating fluorescent dyes or probes (SYBR, RoX, Hex, Vic, Fam, etc.), real-time PCR enables the continuous monitoring and quantification of DNA or RNA amplification in real time. In this method, a DNA or RNA sample is combined with primers, nucleotides, and DNA polymerase, alongside specialized fluorescent components. Several reference or internal control genes are examined for stable gene expression. The real-time PCR instrument, equipped with a thermal cycler, cycles through temperature stages, facilitating denaturation, annealing, and extension steps. Fluorescence emitted during amplification is detected after each cycle, with the intensity directly correlating to the amount of amplified genetic material. This approach provides quantitative data, crucial for applications like gene expression studies, pathogen detection, viral load determination, and genetic variation analysis. Real-time PCR simplifies the analytical process by eliminating the need for post-amplification procedures, thereby streamlining workflows and reducing the potential for contamination or errors. Overall, its capacity to provide precise, sensitive, and automated results has established real-time PCR as an indispensable tool in a multitude of scientific disciplines.

Key words Real-time PCR, Fluorescent dyes, Gene expression, Quantitative data, Reference or internal control genes

1 Introduction

Real-time polymerase chain reaction (PCR) has become an essential tool in the field of plant virology, providing accurate and efficient detection as well as quantification of viral pathogens within plant samples [1]. Real-time PCR uses fluorescence to measure the amount of amplicon produced during each amplification cycle. Instead of measuring the amount of product at the end of the reaction, real-time PCR may quantify amplicon synthesis during the exponential phase of the PCR. Real-time monitoring of the amplicon involves labeling and detecting the increasing product

Aflaq Hamid et al. (eds.), *Detection of Plant Viruses: Advanced Techniques*, Springer Protocols Handbooks,
https://doi.org/10.1007/978-1-0716-4390-7_23,

with a fluorescently tagged substrate during the amplification operation. This approach has numerous advantages over conventional PCR, including increased speed due to reduced cycle number, the absence of post-PCR gel electrophoresis product detection, and the higher sensitivity of the fluorescent dyes employed to detect the amplicon [2].

The heightened sensitivity and specificity of real-time PCR have revolutionized plant virology, allowing for early and precise identification of viral infections [3]. This capability is particularly crucial for distinguishing closely related viral strains, aiding in the targeted implementation of control measure [4]. The quantitative nature of real-time PCR plays a pivotal role in measuring viral nucleic acids precisely within plant tissues [5]. This feature is vital for monitoring changes in viral load over time, providing valuable insights into the dynamics of infection, the interactions between hosts and pathogens, and the overall impact on plant health. Real-time PCR also contributes significantly to studying the genetic diversity of plant viruses [6]. Its use in genotyping and strain differentiation enhances our understanding of the evolutionary dynamics of viral populations, which is essential for predicting disease emergence and developing crop varieties resistant to specific viruses [3]. The scalability of real-time PCR allows for the high-throughput screening of large batches of plant samples [7]. Multiplexing capabilities enable the simultaneous detection of multiple viral species, streamlining diagnostic processes and increasing the efficiency of extensive surveillance programs [8]. In agriculture, real-time PCR data serves as the cornerstone for precision disease management. Information on viral load and strain diversity guides the implementation of targeted control strategies, including the strategic deployment of resistant cultivars, optimized pesticide use, and timely interventions to curb the spread of viral diseases [9].

2 Materials

1. Template (cDNA).
2. SYBR Green Master Mix (2×).
3. Forward primer.
4. Reverse primer.
5. Nuclease-free water.

3 Methods

3.1 Primer Design

The key features of effective real-time RT-PCR primer design are outlined below.

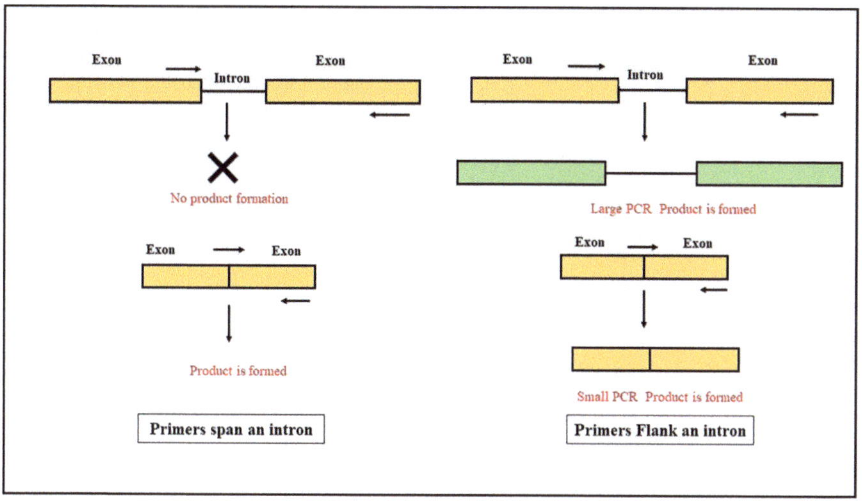

Fig. 1 Designing primer for real-time PCR. (Chap. 23)
Primers should ideally be made that span an intron so that they can only anneal to the cDNA produced from the spliced message RNA as seen on the left. Primers wouldn't work well with any contaminated genomic DNA. Alternatively, a long intron can be flanked by primers such that only cDNA produced from a spliced message RNA would effectively create a PCR result, as indicated on the right

1. Primer pairs should amplify an amplicon of appropriate size (preferably 100–200 bp) (*see* **Note 1**).
2. Primer sequences should span or flank a large intron.
3. Primer binding sites should have minimal secondary structure.
4. The primer's three ends should not contain secondary structure or repetitive sequences.
5. Primer pairs should not considerably complement one another or themselves.
6. Primer pairs should have equivalent GC concentration (40–70%) and similar annealing temperatures (Fig. 1).

3.2 Reaction Setup

1. Thaw, mix, and briefly centrifuge each component before use.
2. Prepare the reaction mixture (minus DNA template) by combining the SYBR Green Master Mix, forward primer, reverse primer and nuclease free water as described below. The template cDNA is added in **step 3**. Vortex briefly to mix.

Components	Volume	Final Concentration
SYBR green master mix (2×)	10 μL	1×
Forward primer (20×)	_____μL	200 nM–1 μM

(continued)

Components	Volume	Final Concentration
Reverse primer (20×)	____μL	200 nM–1 μM
Nuclease-free water	To a final volume of 20 μL	
Final volume	20 μL	

3. Add the appropriate volume of reaction mixture to each PCR tube (*see* **Note 2**).
4. Add the cDNA template to the PCR tube. (*see* **Note 3**).
5. Seal the tubes, and centrifuge briefly to collect all the contents of the tube at the bottom. The samples are ready for thermal cycling. Protect the samples from extended light exposure or elevated temperature (*see* **Note 4**).
6. Include a "no template control" in the qPCR setup for every gene examined to look for potential cross-contamination from sample processing (*see* **Note 5**).
7. For stable gene expression, several reference genes (e.g., *EF1a*, actin, *GAPDH*, etc.) should be examined. Choose a reference gene whose expression pattern remains unchanged during the various testing time points or during treatment [10].

3.3 Thermal Cycling

Always follow the cycling conditions according to the master mix used (*see* **Note 4**).

Step	Temperature	Time	Cycle
Initial denaturation	95 °C	2 min	1
Denaturation	95 °C	30 s	35
Annealing	60 °C	30 s	35
Extension	72 °C	60 s	35

(*Use the instrument optical setting established for SYBR Green Master Mix*)

3.4 Determine Ct Values

1. Utilize the relevant software (e.g., QuantStudio from Applied Biosystems, CFX Manager from Bio-Rad, or others) provided by the qPCR equipment manufacturer to examine the fluorescence data following the qPCR run.
2. To find the Ct values for each sample, either manually or automatically set the baseline and threshold.

3.5 Calculate Relative Expression: ΔΔ Ct Method

1. Select appropriate housekeeping genes as reference genes for normalization, ensuring their expression remains consistent across all samples.

2. Compute the ΔCt for individual samples by deducting the Ct value of the chosen reference gene from the Ct value of the target gene.
3. Determine the ΔΔCt by subtracting the ΔCt of a reference sample (e.g., control) from the ΔCt of each experimental sample.
4. Evaluate relative expression levels using the formula: 2^(-ΔΔ Ct) [11].

3.6 Data Interpretation

1. Analyze results based on Ct values.
2. Lower Ct values indicate higher initial concentrations of the target nucleic acid.
3. Compare Ct values between experimental samples and controls. Higher Ct values in experimental samples may suggest lower target concentrations.

3.7 Verify Specificity Using Melting Curve Analysis

1. Conduct a melting curve analysis promptly following the qPCR run to evaluate the specificity of the amplification (Fig. 2).
2. The melting curve shows the dissociation of the PCR product as the temperature increases. A single peak indicates specific amplification (Fig. 3).
3. Verify the existence of primer dimers or nonspecific products, as they could lead to extra peaks or irregularities in the melting curves.

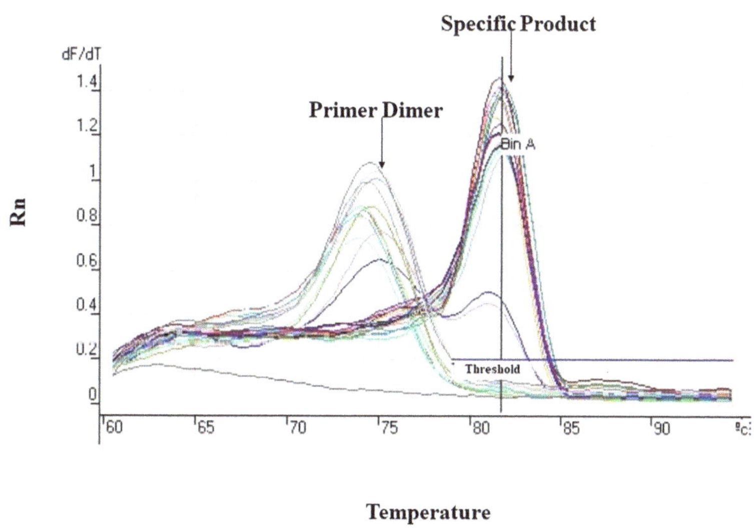

Fig. 2 Real-time PCR melt curve profile. (Chap. 23)

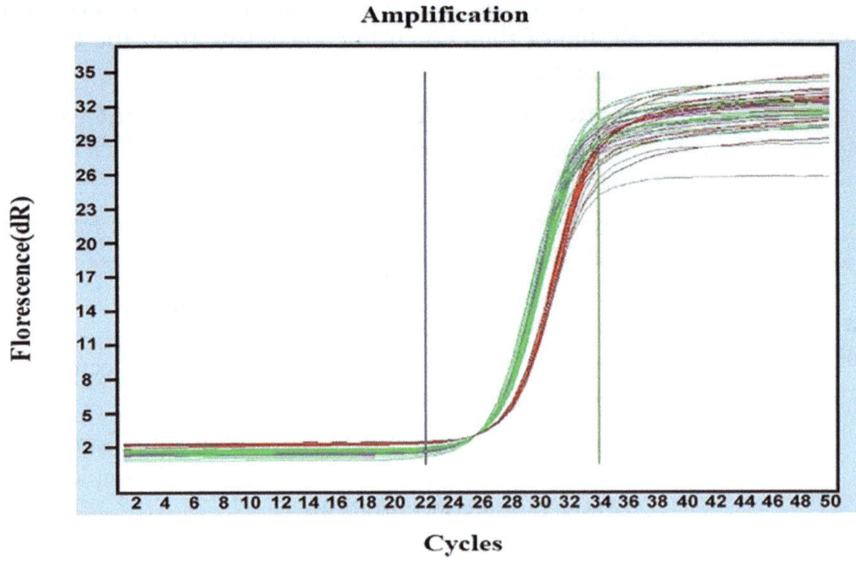

Fig. 3 Real-time PCR amplification profile. (Chap. 23)

3.8 Draw Conclusions

1. Evaluate the relative expression data that the ΔΔCt method generated.
2. Examine how the target gene is expressed differently in experimental samples compared to reference or control samples.
3. Consider statistical analysis to determine the significance of any observed differences.

4 Notes

1. Design primers and probes specifically for the target sequence, avoiding secondary structures and primer-dimers, and validate them for efficiency and specificity.
2. To minimize pipetting errors and variability, use master mixes. Keep all reagents on ice, thaw completely before use, mix thoroughly but gently, and use freshly prepared reagents or aliquots stored properly to avoid repeated freeze-thaw cycles.
3. Ensure the use of high-quality, accurately measured nucleic acid samples along with RNase-free/DNase-free reagents and consumables to prevent contamination and degradation, as contaminants can interfere with the PCR.
4. Optimize cycling conditions, encompassing denaturation, annealing, and extension temperatures and durations, and employ a thermal cycler featuring precise temperature regulation.

5. Ensure assay reliability by consistently incorporating positive and negative controls, running technical replicates to gauge result consistency, and employing suitable threshold settings and baselines in the qPCR software for precise quantification.

References

1. Bustin SA (2000) Absolute quantification of mRNA using real-time reverse transcription polymerase chain reaction assays. J Mol Endocrinol 25:169–193
2. Fraga D, Meulia T, Fenster S (2014) Real-time PCR. Curr Protoc Essent Lab Tech 8:10–13
3. Mackay IM, Arden KE, Nitsche A (2002) Real-time PCR in virology. Nucleic Acids Res 30: 1292–1305
4. Bustin SA, Benes V, Garson JA, Hellemans J, Huggett J, Kubista M, Mueller R, Nolan T, Pfaffl MW, Shipley GL, Vandesompele J, Wittwer CT (2009) The MIQE guidelines: minimum information for publication of quantitative real-time PCR experiments. Clin Chem 55:611–622
5. Pfaffl MW (2001) A new mathematical model for relative quantification in real-time RT-PCR. Nucleic Acids Res 29
6. Guescini M, Sisti D, Rocchi MB, Stocchi L, Stocchi V (2008) A new real-time PCR method to overcome significant quantitative inaccuracy due to slight amplification inhibition. BMC Bioinformatics 9:326
7. Bustin SA, Mueller R (2005) Quantification of mRNA using real-time reverse transcription PCR (RT-PCR): trends and problems. J Mol Endocrinol 34:597–601
8. Nolan T, Hands RE, Bustin SA (2006) Quantification of mRNA using real-time RT-PCR. Nat Protoc 1:1559–1582
9. Pabinger S, Rödiger S, Kriegner A, Vierlinger K, Weinhäusel A (2014) A survey of tools for the analysis of quantitative PCR (qPCR) data. Biomol Detect Quantif 1:23–33
10. Radonic A, Thulke S, Mackay IM, Landt O, Siegert W, Nitsche A (2004) Guideline to reference gene selection for quantitative real-time PCR. Biochem Biophys Res Commun 313: 856–862
11. Livak KJ, Schmittgen TD (2001) Analysis of relative gene expression data using real-time quantitative PCR and the 2− ΔΔCT method. Methods 25:402–408

Chapter 24

Multiplex PCR in Plant Virus Detection

Sumiah Wani, Shahjahan Rashid, Nulevino Iralu, Sajad Un Nabi, and Aflaq Hamid

Abstract

Multiplex PCR is a powerful technique for viral detection that allows the simultaneous amplification of multiple DNA or RNA targets in a single reaction. Multiplex PCR, which simultaneously amplifies more than one locus in the same reaction, is an important clinical and research laboratory screening tool. This method is particularly valuable in plant virology as it enables the detection of several viruses or viral strains within the same sample, saving time and resources compared to performing individual PCR assays. By using multiple sets of primers specific to different viral sequences, multiplex PCR provides a comprehensive view of the viral population, which is crucial for diagnosing mixed infections and managing complex viral outbreaks. Additionally, this technique enhances diagnostic efficiency and accuracy, facilitates rapid screening of large numbers of samples, and helps in the development of more effective control measures and resistant plant varieties. The relative concentrations of the primers at the various loci, the concentration of the PCR buffer, the cycle temperatures, and the balance between the magnesium chloride and deoxynucleotide concentrations are all critical for a successful multiplex PCR experiment. Because the reaction has a greater number of primers, appropriate primer design is the key to the success of multiplex PCR reaction. Multiplex PCR is a key technique for the simultaneous detection for co-infection of viruses in one plant.

Key words cDNA, Reaction, Primers, Mastermix

1 Introduction

In the context of plant viruses, multiplex PCR plays a critical role in enhancing diagnostic efficiency and accuracy. This technique allows for the simultaneous detection of multiple plant viruses or viral strains from a single sample, which is particularly useful given the diverse array of viruses that can affect crops [1]. By combining multiple virus-specific primers in one reaction, multiplex PCR provides a comprehensive screening tool that can identify several pathogens at once, reducing the need for multiple individual tests. The efficiency of multiplex PCR not only speeds up the diagnostic process but also conserves resources [2]. With a single assay capable of detecting multiple viruses, there is less need for

Aflaq Hamid et al. (eds.), *Detection of Plant Viruses: Advanced Techniques*, Springer Protocols Handbooks,
https://doi.org/10.1007/978-1-0716-4390-7_24,

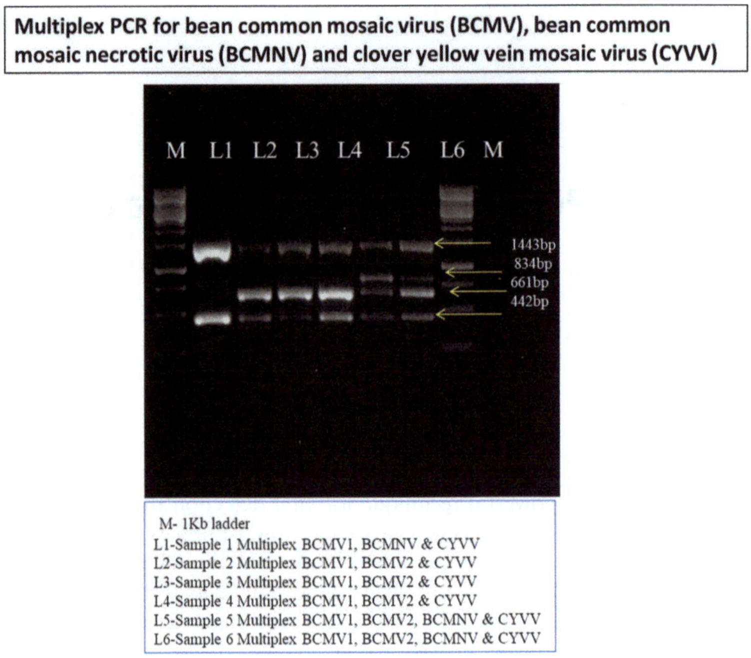

Fig. 1 Multiplex PCR for BCMV, BCMNV, and CYVV

extensive reagent use and fewer samples required, thereby lowering the costs associated with virus detection and simplifying the workflow. This is particularly advantageous in agricultural settings where timely and cost-effective diagnostics are essential for managing plant health.

Moreover, the sensitivity and specificity of multiplex PCR are crucial for accurate plant virus detection. The technique can detect low levels of viral nucleic acids, which is important for early identification of infections before they spread widely. By using specific primers for each target virus, multiplex PCR minimizes the risk of false positives and negatives, ensuring reliable results that can guide effective management strategies. In addition, multiplex PCR is valuable for monitoring and managing plant diseases on a broader scale [3–5]. It supports the rapid identification of viral outbreaks, helping to implement targeted control measures and reduce the spread of diseases. In research and breeding programs, this technique aids in the study of viral diversity, evolution, and resistance, contributing to the development of resistant plant varieties and improved disease management practices [6].

Overall, multiplex PCR significantly enhances the ability to detect and manage plant viruses, providing a powerful tool for both practical agricultural applications and scientific research (Fig. 1).

2 Materials

- Cdna.
- Gotaq Green.
- Primers (*see* **Note 1**).
- Nuclease-free water.
- Thermocycler.

3 Methods

(a) cDNA synthesis: Synthesis of cDNA using required chemicals from cDNA synthesis kit and standard protocol (*see* **Note 2**). Tubes containing the reaction mixture should be incubated for 1 h at 37 °C followed by 15 min at 72 °C in a thermocycler (*see* **Note 3**).

(b) PCR Master mix preparation.

- The quantity of master mix will depend upon the number of samples to be processed.

Standard protocol 25 uL for 1× reaction	
GoTaq green	12.5 uL
Reverse primer	1 uL
Forward primer	1 uL
Nuclease-free water	8.5 uL
Total	23 uL

- Take a microtube and add GoTaq Green, forward primer, reverse primer, and nuclease-free water. Keep the master mix in ice (*see* **Note 4**).
- Take 2ul of cDNA template in PCR tube and add 1x reaction mixture to each tube from the master mix tube (*see* **Note 5**). Keep in ice to avoid degradation.

(a) PCR program: (*see* **Note 6**)

S. No	Stage	Temperature (°C)	Time
1.	Initial denaturation	95	5 min
2.	Denaturation	95	1 min
3.	Annealing	50–60 (based on primers)	1.5 min
4.	Extension	72	1 min
5.	Final extension	72	7 min

Stage 2 to 4 should be repeated for 30 cycles (*see* **Note 7**).

(b) After setting the PCR program, start the program. Once the program is over, keep the PCR tubes in ice and load in 1% agarose gel then visualize under gel doc (*see* **Note 8**).

4 Notes

1. Design Specific Primers: Ensure primers are specific to their targets to avoid nonspecific amplification. Use software tools for designing primers that minimize cross-reactivity and self-dimerization.
2. Use high-quality reagents and check for the expiration dates of enzymes, buffers, and other components. Store reagents as recommended to maintain their integrity.
3. Adjust annealing temperatures and concentrations of reagents to suit the primer and targets used. Validate conditions through preliminary testing.
4. Prevent Cross-Contamination and use separate areas and equipments for preparing PCR reactions and for post-PCR analysis. Employ dedicated pipettes and filter tips and work in a clean environment with regular decontamination using UV light or chemical disinfectants.
5. Also include both positive controls (with known target) and negative controls (without DNA) to verify that the PCR reaction is working and to identify potential contamination.
6. Follow recommended thermal cycling protocols and ensure that the PCR machine is properly calibrated. Use a gradient PCR if necessary to optimize annealing temperatures. Also Test the multiplex PCR with various combinations of primers to ensure they work together without interfering with each other. Adjust concentrations as needed.
7. Check for Amplification Efficiency: Regularly assess the efficiency of the amplification of each target to ensure that the multiplex assay is functioning as intended.
8. Document and Analyze Results Carefully: Carefully interpret the PCR results and analyze gel electrophoresis or other detection methods to distinguish between different target sequences and to identify any potential issues.

References

1. Inifro EM, Ashshi AM, Cooper RJ, Klapper PE (2000) Multiplex PCR: optimization and application in diagnostic virology. Clin Microbiol Rev 13(4):559–570
2. Mehetre GT, Leo VV, Singh G, Sorokan A, Maksimov I, Yadav MK, Upadhyaya K, Hashem A, Alsaleh AN, Dawoud TM, Almaary KS (2021) Current developments and

challenges in plant viral diagnostics: a systematic review. Viruses 13(3):412

3. Markoulatos P, Siafakas N, Moncany M (2002) Multiplex polymerase chain reaction: a practical approach. J Clin Lab Anal 16(1):47
4. Sint D, Raso L, Traugott M (2012) Advances in multiplex PCR: balancing primer efficiencies and improving detection success. Methods Ecol Evol 3(5):898–905
5. Wani S, Saleem S, Nabi SU, Ali G, Paddar BA, Hamid A (2021) Distribution and molecular characterization of potato virus Y (PVY) strains infecting potato (Solanum tuberosum) crop in Kashmir (India). VirusDisease 32:784–788
6. Huang A, Gu P, Wang Y, Li J, Yang Z, Yi L (2023) Simultaneous detection and differentiation of four viruses in passion fruit plants by a multiplex RT-PCR. Trop Plant Pathol 48(1): 23–29

Chapter 25

Colony PCR in Plant Virology

Mehreena Farooq Rather, Sumiah Wani, Wani Farhana, Nulevino Iralu, Shahjahan Rashid, and Aflaq Hamid

Abstract

Molecular cloning necessitates a method for screening colonies to confirm the presence of a recombinant insert. Traditionally, this has been achieved through restriction enzyme digestion; however, colony polymerase chain reaction presents a more efficient and cost-effective alternative. It allows direct amplification of DNA from cellular colonies without prior DNA purification, thereby facilitating rapid verification of the expected DNA sequence. This technique significantly reduces the time and expense associated with traditional methods, providing a streamlined approach for sequence verification in molecular cloning studies.

Key words Colony PCR, DNA, Bacterial colony

1 Introduction

Colony polymerase chain reaction (PCR) is a methodological approach used to amplify a specific region of the target genetic construct, enabling rapid screening of yeast or bacterial colony that have proliferated on selective media following transformation [1]. This method is utilized to rapidly assess bacterial colonies formed after the transformation stage of a molecular cloning protocol, to verify that the desired genetic construct is present or to amplify a portion of the construct [2, 3].

The processes of plasmid DNA extraction, restriction enzyme digestion, and gel fragment analysis, which are typically labor-intensive and time-consuming, are streamlined using this method. It is routinely used to verify the presence of insert DNA in applications involving gene transfer, cloning, and genetic engineering. Different variants of colony PCR have been devised for screening of fungi, yeasts, bacteria, and, more recently, specific microalgae. These efficient and time-saving methods, typically involve re-suspending the cells in specialized buffers, followed by a critical

Aflaq Hamid et al. (eds.), *Detection of Plant Viruses: Advanced Techniques*, Springer Protocols Handbooks,
https://doi.org/10.1007/978-1-0716-4390-7_25,

thermal lysis step that facilitates cellular disruption. In case of yeast (*Saccharomyces cerevisiae*), the cells are lysed by boiling in 0.2% SDS at 98 °C for 5–10 min to release genomic DNA [4]. The effectiveness of colony PCR relies on the use of specifically designed primers that hybridize to their complementary sequences on the plasmid DNA thus enabling the DNA polymerase enzyme to selectively amplify the desired region. These oligonucleotides are engineered to anneal to sequences flanking the target region within the inserted DNA, for high-throughput gene screening, such as in the PCR amplification of small-subunit ribosomal DNA sequences from bacterial species [5]. PCR of genomic DNA directly from bacterial colonies is a routine requirement in molecular biology laboratories and is an invaluable tool for researchers in molecular cloning and genetic engineering.

1.1 Equipment Used in Colony PCR

1. Petri dishes, containing colonies to be tested (*see* **Note 1**).
2. PCR Thermocycler.
3. Benchtop microcentrifuge.
4. Gel electrophoresis equipment.
5. Vortex mixer.
6. Microwave oven.
7. Micropipette tips (*see* **Note 2**).
8. Transilluminator, or other UV light source.
9. Micropipette tip.
10. 0.2-mL thin-walled PCR tubes.

2 Materials

1. 10× PCR buffer (supplied with DNA polymerase).
2. 2.5 mM (each) dNTP mix (supplied with DNA polymerase).
3. 30 μM oligonucleotide primer 1.
4. 30 μM oligonucleotide primer 2.
5. 5 U/μL Taq DNA polymerase, heat-stable (e.g., TaKaRa Ex Taq DNA polymerase.
6. Takara or Platinum Taq DNA polymerase, Thermo Fisher Scientific.
7. Distilled H_2O, sterile.
8. $MgCl_2$ (if required).
9. NaOH.
10. Gel electrophoresis loading buffer.

11. 200-μL thin-walled PCR tubes (or other appropriate tubes for thermal cycler).
12. Toothpicks.
13. Purified water.
14. Thermal cycler.

2.1 Important Points for Colony PCR

1. *Template Preparation.*

   Utilize a sterile flat toothpick or pipette tip to isolate a single colony, and suspend it in a minimal volume of sterile water. Repeat this procedure for 3–10 colonies in total to test, depending on the number of background colonies on your no-ligation control plate. The more background, the more colonies you will need to screen (*see* **Notes 3** and **4**).
2. *Preservation of Clones for Subsequent Culture.*

   At this stage, it is advisable to retain your clones for future applications. Several strategies can be employed for this purpose. If colony PCR analysis is completed on the same day, the residual bacteria-water suspension can be utilized to initiate cultures of positive clones. For longer-term storage, streak the colonies onto an LB agar plate, which can later be used to inoculate liquid cultures. Alternatively, initiate small overnight liquid cultures with the selected clones, and subsequently perform mini-prep on the positive cultures. Regardless of the chosen method, ensure the use of the appropriate antibiotic for selective pressure.
3. *Bacterial Lysis and PCR Setup.*

   The residual bacteria-water suspension will function as the template for the PCR. Bacterial lysis can be achieved by briefly boiling the sample before use, or by directly incorporating a small volume of the suspension into the PCR. During the initial denaturation step of the PCR cycle, the bacteria will undergo lysis, releasing the plasmid DNA (*see* **Note 5**).
4. *Control.*

   Implementing appropriate controls is crucial for the success of an experiment. Optimal controls for colony PCR include the backbone vector with and without an insert, which are the same controls used to validate the efficacy of the colony PCR primers. These controls provide reference points when analyzing PCR products via gel electrophoresis to ascertain the presence of the insert in the colonies. Additionally, they serve as controls for the PCR itself. Including a no-template control PCR is also advisable to detect any potential contamination.

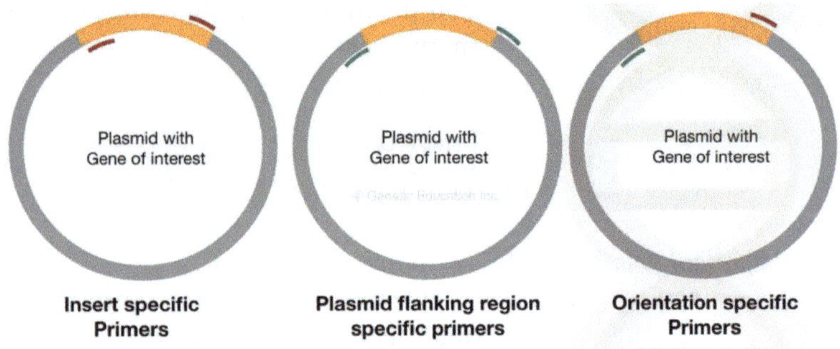

Fig. 1 Different primer design strategies for colony PCR. (Chap. 25)

2.2 Preparation of Colony PCR

The success of colony PCR depends on the design of the proper primers.

Primers designed for colony PCR may provide details about the following, depending on what they are intended to target:

- The genetic insert is present, and primers are made to target it rather than the plasmid (insert-specific primers).
- Primers are made to flank the insert based on the insert size (backbone-specific primers).
- The insert is oriented so that one primer targets the insert and the other primer targets the plasmid (orientation-specific primers) (Fig. 1).

2.3 Interpretation of PCR Results

Positive clones will exhibit a band upon gel electrophoresis analysis of the PCR product when using insert-specific primers, whereas negative clones will display no band. Employing backbone-specific primers, positive clones generate larger-sized products compared to negative clones. Furthermore, orientation-specific primers can determine the correct orientation of the insert, yielding the same band or no band outcome as insert-specific primers.

1 Insert specific primers	2 Backbone specific primers	3 Orientation specific primers

3 Methods

3.1 Basic Protocol for Bacteria

1. Prepare a master mix for PCR amplification (e.g., 100 μL for ten tubes, 10 μL each):
   - 10 μL 10× PCR buffer containing 15 mM $MgCl_2$.
   - 8 μL of 2.5 mM (each) dNTP mix..
   - 5 μL of 30 μM primer 1.
   - 5 μL of 30 μM primer 2.

- 1 μL of 5 U/ml heat-stable Taq DNA polymerase.
- 71 μL sterile distilled water.

2. Dispense master mix into PCR tubes.
3. Take a small portion of a bacterial colony straight out of the plate to be examined and use a fresh, sterile toothpick for each colony (*see* **Note 6**).
4. In a PCR tube containing the master mix, insert a toothpick. Incorporate bacteria into the PCR mixture by scraping it in. Simply twist the toothpick between your thumb and forefinger after pushing its tip to the bottom of the tube. As soon as possible, remove the toothpick.
5. Perform PCR at conditions appropriate for the DNA being examined, using the following amplification cycles as a guideline (*see* **Note 7**):

Initial step:	5 min	94 °C (bacterial lysis/denaturation)
25–35 cycles:	30 s	94 °C (denaturation)
	30 s	50–55 °C (annealing)
	1 min	68–72 °C (extension)
Final extension:	5 min	72 °C.

6. Incorporate 5–10 μL of each reaction mixture with 5 μL of gel electrophoresis loading buffer.
7. To electrophorese the PCR result, use either an agarose gel [6] or a non-denaturing, continuous polyacrylamide gel [7] based on its size.
8. To visualize PCR results, stain the gel with ethidium bromide and expose it to UV light (*see* **Note 8**).
9. Ascertain whether the insert is present and/or how it is oriented.

3.2 Alternate Protocol for Bacteria

In basic protocol, the initial heating stage of 94 °C is necessary to lyse bacteria and release DNA. Certain bacteria, such as Gram-positive species that have thick walls, might not lyse effectively in those circumstances. Additionally, agar and other materials in the sample, as well as an excessive amount of bacteria collected, may hinder PCR. As a result, using the Alternative Protocol may also boost the effectiveness of *Escherichia coli* colony PCR. But it does lengthen the process by around 30 min.

3.3 Modifications in Basic Protocol

Additional Materials

1. Distilled water.
2. 500-μL microcentrifuge tubes.

3. Water bath (boiling) or 100 °C heat block.
4. Microcentrifuge.

3.3.1 Procedure

1. Dispense 50 μL distilled water into microcentrifuge tubes (One tube per bacterial colony).
2. Follow **steps 1** through **3** in Basic Protocol.
3. Place a toothpick into a microcentrifuge tube. Scrape bacteria into the water, as above.
4. Cap tubes, and place in boiling water bath or 100 °C heat block for 10 min.
5. Pellet bacterial lysates in a microcentrifuge at top speed.
6. Mix 2 μL supernatant with 8 μL PCR mix.

   Reduce the amount of distilled water in the master mix to account for the 2 μL of DNA that is added [8].
7. Follow **steps 5–9**, Basic Protocol.

3.3.2 Protocol for Bacteria, Yeasts, and Microalgae

1. Pick up cells from a colony using a sterile pipette tip and then resuspend into 20 μL of each of the different lysis buffers (*see* **Note 9**).
2. Vortex the mixture for 10 s to lyse and disrupt cells and then incubate for 5 min at 98 °C if necessary.
3. Collect the supernatant after microcentrifugation of lysate and dilute with distilled water at a 1:5 ~ 1:20 dilution ratio, followed by PCR analysis.
4. Analyze the extraction efficiency of genomic DNA by various buffers using PCR amplification of the 16 s rRNA for bacteria and the 18S rRNA for yeasts and microalgae (*see* **Note 10**).

   Carry out PCR in 20 μL of 1× high fidelity buffer, 2 pmoL of each primer, 4 μmoL of each dNTP, 1–2 μL of diluted genomic DNA templates extracted previously, and 1 unit of DNA polymerase with high fidelity for 30 cycles using thermal cycler [4].
5. Each cycle consists of the following:

30 cycles:	10 s 30 s 45 s	98 °C (denaturation) 56 °C (annealing) 72 °C (extension)
Final extension:	5 min	72 °C

6. Analyze the PCR products by 0.8% agarose gel electrophoresis.

4 Notes

1. This procedure involves opening agar plates containing genetically modified bacteria. It is not good practice to open agar plates that contain unknown microbes after incubation, so only uncontaminated agar plates (containing only the expected red and white *E. coli* colonies) should be used.
2. Plates and all other disposable materials used in this laboratory must be sterilized before disposal.
3. It is important that the toothpicks have not been used before. Previously used toothpicks, even if they have been autoclaved, may carry DNA from earlier transfers and give false positive results.
4. Be certain to touch just one colony and avoid colonies very close to each other, otherwise this could contaminate the results.
5. Adjust the amount of bacteria added to the PCR; consider transferring each colony to a gridded plate, incubating overnight, then using the transferred smear as a template for PCR.
6. Best efficiency is generally achieved when the bacteria are barely visible on the toothpick tip.
7. If multiple PCR products are obtained it is because of nonspecific binding of primers to DNA, in that case, increase annealing temperature 3–5 °C below primers Tm.
8. If the colony PCR product is of appropriate size but not the desired DNA sequence, restreak for individual colonies and repeat the PCR screen before sequencing.
9. Avoid using too much-starting material. It's important when adding the lysed cells to the PCR mix that too much material isn't added to the DNA template as this can result in nonspecific amplification.
10. Include controls. Both positive and negative controls should be included in colony PCR. A good positive control would be any DNA that you are sure should amplify using the primers selected; this helps to ensure that the PCR "worked," i.e., that the reagents used haven't expired, etc. It is also recommended that a negative control is included in one reaction that hasn't been exposed to a bacterial colony to ensure that your PCR reagents are clean.

4.1 Application of Colony PCR

1. Colony PCR is used by researchers in CRISPR-CAS9-like investigations, gene transfer, gene therapy, gene cloning, and genetic modification.
2. Colony PCR is mostly used to assess if yeast and bacteria are ligating correctly.

4.2 Advantages of Colony PCR

1. The technique is rapid and cost-effective.
2. It is more accurate and specific.
3. DNA extraction and plasmid purification aren't required like conventional PCR.
4. It is possible to avoid laborious, expensive, and time-consuming procedures like restriction digestion. Only a basic colony PCR can function effectively and produce the desired results.
5. It can be effectively used not only to identify clones with an insertion or deletion but also to determine the orientation of a DNA insertion, which may be important for proper transcription and translation of the insert.
6. The method can be used to amplify a desired DNA fragment for subsequent sequencing or cloning.
7. While colony PCR is most commonly used in the laboratory to screen transformed *E. coli*, DNA sequences from virtually any bacterial species can be detected and/or isolated regardless of location on the chromosome or a plasmid.

4.3 Disadvantages of Colony PCR

1. We must rely on sequencing because it is unable to provide us with sequence information.
2. False-positive outcomes are highly likely.

References

1. Bergkessel M, Guthrie C (2013) Colony PCR. Methods Enzymol 529:299–309
2. Azevedo F, Pereira H, Johansson B (2017) Colony PCR. PCR: methods and protocols. Methods Mol Biol 1620:129–139
3. Güssow D, Clackson T (1989) Direct clone characterization from plaques and colonies by the polymerase chain reaction. Nucleic Acids Res 17(10):4000
4. Packeiser H, Lim C, Balagurunathan B, Wu J, Zhao H (2013) An extremely simple and effective colony PCR procedure for bacteria, yeasts, and microalgae. Appl Biochem Biotechnol 169: 695–700
5. Janda JM, Abbott SL (2007) 16S rRNA gene sequencing for bacterial identification in the diagnostic laboratory: pluses, perils, and pitfalls. J Clin Microbiol 45(9):2761–2764
6. Voytas D (2000) Agarose gel electrophoresis. Curr Protoc Mol Biol 51(1):2–5
7. Chory J, Pollard JD Jr (1999) Separation of small DNA fragments by conventional gel electrophoresis. Curr Protoc Mol Biol 47(1):2–7
8. Woodman ME, Savage CR, Arnold WK, Stevenson B (2016) Direct PCR of intact bacteria (colony PCR). Curr Protoc Microbiol 42(1):A-3D

Chapter 26

Application of Digital PCR (dPCR) in Plant Virus Detection

Insha Mehraj, Insha Mehraj, Nulevino Iralu, Fizalah Kawoosa, Sumiah Wani, Sahar Saleem, and Aflaq Hamid

Abstract

Pathogens present a significant risk to both food safety and public health. Traditional pathogen detection methods are increasingly inadequate for meeting the demands of modern diagnostics and control. Digital PCR (dPCR) has gained substantial devotion due to its capability to precisely quantify pathogens, offering advantages such as high selectivity, ease of use, accuracy, and speed. Droplet digital PCR (ddPCR) is a highly sensitive molecular biology tool utilized to quantify nucleic acids with absolute precision. It fractionates a PCR mixture into thousands of nanoliter-sized droplets using water-oil emulsion technique and each droplet acts as a separate individual micro-reaction during PCR, thus increasing amplification efficiency manifold. It follows a simple workflow wherein portioning is the primary step, followed by PCR amplification in a compatible thermal cycler and finally detection and quantification using software. Compared to traditional PCR methods, it provides increased sensitivity, reduced variability, and enhanced accuracy.

Key words ddPCR, Water-oil emulsion, Absolute quantification, Sensitivity

1 Introduction

In recent years, polymerase chain reaction (PCR) techniques have been advanced to provide high specificity, precision, reproducibility, and sensitivity in pathogen detection [1]. As PCR has evolved, various derived techniques have emerged, including multiplex allele PCR, multiplex PCR, inverse PCR, nested PCR, fluorescence immune PCR (IPCR), quantitative PCR (qPCR), anchored PCR, long fragment PCR, asymmetric PCR, and digital PCR (dPCR). These techniques are all utilized for the amplification of target nucleic acids and recognition. While qPCR is widely used for pathogen detection, it is limited by its reliance on cycle threshold (Ct) measurements, which provide relative quantification. Consequently, there is a pressing prerequisite for a more rapid, accurate, and sensitive detection method that offers a lower LOD (limit of detection) for pathogens and is less affected by inhibitors present in complex samples [2].

Aflaq Hamid et al. (eds.), *Detection of Plant Viruses: Advanced Techniques*, Springer Protocols Handbooks,
https://doi.org/10.1007/978-1-0716-4390-7_26,

Droplet digital PCR (ddPCR) is a recently emerging technique used in detection and diagnostic methods, also applied to use in case of plant viruses. The term "digital" refers to the binary nature of signal detection (e.g., on or off, coagulated or uncoagulated, activated or inactive, fluorescent or non-fluorescent) in individual entities. In digital PCR (dPCR), DNA templates are first diluted to a specific concentration typically ranging from 100 to 1000 copies per microliter, which statistically ensures that each reaction chamber contains either one or zero molecules before amplification. However, even with sufficient dilution, there is always a statistical possibility of more than one molecule being present in a reaction. To account for this, maximum dPCR instruments use the Poisson distribution [3]. Recently, novel applications have proposed using negative binomial distributions, as the Poisson distribution assumes no variation in partition sizes, which is only accurate in a theoretical context [4]. Amplification products are typically detected using chambers and fluorescent probes containing a single template molecule that will produce positive signals. In rare cases where two or more target molecules are present, the resulting unusual fluorescence intensity can be identified and excluded from analysis. As a result, numerous chambers with a positive signal provide a digital readout, enabling absolute target quantification and mitigating the problem of amplification bias [5].Compared to traditional PCR technologies, digital PCR (dPCR) offers more reproducible, sensitive, precise target quantification. In dPCR, the nucleic acid sample is divided into numerous chambers or compartments (e.g., droplets or microwells), which are independently assessed after amplification. This approach ensures that the outcome is not influenced by deviations in amplification efficiency. Additionally, by evaluating each compartment separately, dPCR reduces background noise and increases the signal-to-noise ratio, thereby notably enhancing detection sensitivity [6]. Currently (dPCR) has garnered noteworthy research attention, as evidenced by the growing number of articles, which primarily explore the concept and advancement of dPCR [7]. Although, some reviews have been conducted on the various commercial dPCR platforms [8]. It was introduced in 2011 and is advanced over conventional PCR with increased efficiency. This is based on water-oil emulsion technology. Unlike conventional PCR, as the name suggests, in droplet digital PCR, a sample is fractionated into 20,000 droplets and each droplet is subjected to PCR amplification of the template molecules, thus increasing the quantity of target molecule in the sample manifold. The droplet represents essentially the same function as individual test tubes in a plate in which PCR takes place, however, in a much smaller format. The immense sample partitioning is the distinguished aspect of ddPCR technique. This technology provides absolute quantification, eliminating the need for standard curves and normalization, and generates straightforward outcomes like "YES" or "NO." This

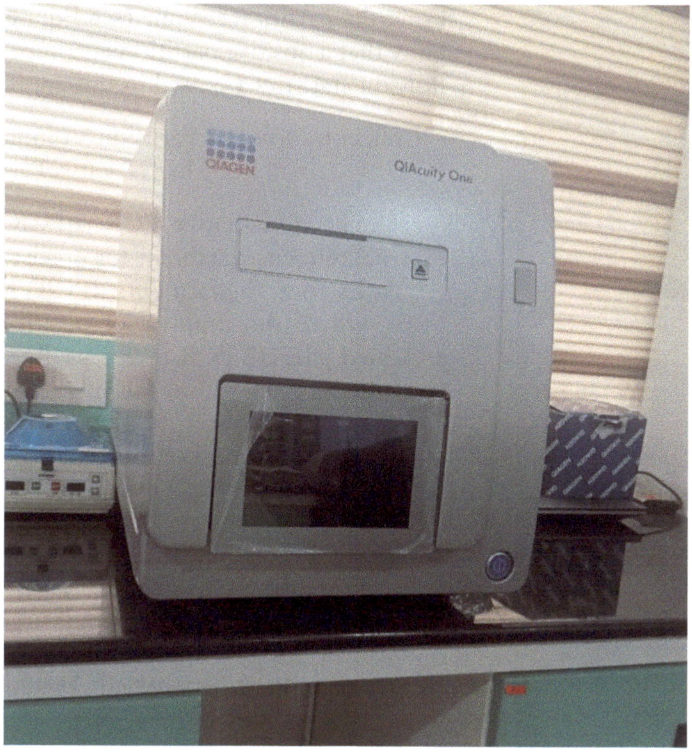

Fig. 1 Digital PCR. (Chap. 26)

technology is beneficial in cases where virus titers are low and show uneven distribution in infected plants. This was proven in the detection of Grapevine red blotch-associated virus (GRBaV) [9]. At the same time, sample requirement and reagent volumes are smaller as compared to other commercially available digital PCR systems, which in turn reduces cost and preserves precious samples. This method is defined by its precision, accuracy, robustness, sensitivity, and scalability over earlier digital PCR techniques (Fig. 1).

1.1 Principle of Digital PCR

Digital PCR (dPCR) functions similarly to quantitative PCR (qPCR) by using the same thermal cycling and chemical parameters to amplify a particular nucleic acid sequence with primers. One of two methods is used to measure the amplification: nonspecific double-stranded DNA-binding dye or fluorescently tagged specific probes.

The key difference in dPCR is that the reaction is divided into 100–10,000 of low-volume partitions. These partitions can either be fixed chambers within a cartridge or droplets within an oil emulsion or in both scenarios, the limiting dilution causes some partitions to contain a target molecule, while others do not. The original initial concentration of the substance may be determined by determining the quantity of positive and negative partitions using the partition volume (see Eq. (1)). To account for the possibility that a positive partition encompasses several target molecules, Poisson statistics are used in this approach.

$$\text{Concentration} = \lambda \text{ Partition Volume; where } \lambda = -\ln\ (\text{NegativePartitions TotalPartitions}) \quad (1)$$

Unlike quantitative PCR (qPCR), dPCR does not need an external reference calibrant for precise quantification.

At its core, digital PCR (dPCR) relies on end-point measurement, where the number of positive and negative partitions is reckoned after all PCR cycles are finished. However, some dPCR equipments available on the market provide real-time dPCR capabilities. These devices monitor the fluorescence in each partition after every cycle and use this data to generate an amplification arc for each partition [10].

1.2 History of Digital PCR

Digital PCR (dPCR) in fact predates quantitative PCR (qPCR). During the late 1980s to early 1990s, various research teams used techniques referred to as "single molecule PCR" and "limiting dilution PCR." In contrast, "Kinetic PCR analysis"was first used to describe qPCR in 1993. In 1999, the acronym "digital PCR" was coined by Bert Vogelstein and Kenneth Kinzler [11].

The foremost commercial digital PCR (dPCR) apparatus, the BiomarkHD (Fluidigm), was introduced in 2006. Since then, other manufacturers have developed dPCR equipments [12]. Over time, dPCR has become more user-friendly and cost-effective, leading to its growing application across various fields in the biosciences. This includes its applications in the study of infectious diseases [2], cancer research [13], genetically modified organisms [14], and inherited diseases [15].

Currently, there is a range of dPCR equipments available, offering various options for factors such as dynamic range, dead volume, partition size, partition type, and number of detection channels [16]. Equipment having numerous detection channels enables simultaneous measurement of multiple targets, which is beneficial for applications with limited sample volumes, like liquid biopsies. However, multiplexing need to be carefully optimized to prevent issues like interference between primer/probe sets and spectral "bleed-through" between channels [17, 18]. The dynamic range and throughput of the instruments can differ significantly.

Instruments with more partitions offer a wider dynamic range, although their sample throughput is frequently lower. So, investigators need to balance these factors when choosing a dPCR instrument. Additionally, some instruments require proprietary reagents, while others can use various commercially available PCR reagents. In cases where non-proprietary reagents are permitted, some laboratories have observed improved outcomes with specific mastermixes [19]. End-users should prudently evaluate their specific needs when choosing a dPCR instrument.

Some studies have compared findings from different dPCR equipments used in different labs as well as within the same lab. For example, [20] demonstrated that, after accounting for biases in partition volume, results from different instruments were comparable when employing a certified reference standard. Similarly, [21] conducted interlaboratory research with isolated DNA and the whole virus, finding that variations between equipment's within a single lab were similar to those between the same equipment in other labs.

1.3 Devices for dPCR

dPCR can be categorized based on the technique used to separate the reaction mixture into two types: chip dPCR (cdPCR) and droplet dPCR (ddPCR).

1.3.1 cdPCR

In cdPCR (chip dPCR), the reaction blend is dispersed into reaction chambers of a nanoliter size using microfluidic technology.

After the reaction cycles, fluorescence is captured using an inverted endoscopy and imaging device. The target sequence's copy number is then counted with imaging device. cDPCR can utilize various technologies, including arrayed lipid bilayer chamber systems (ALBiC) integrated fluidic circuits, and nanoliter self-priming compartmentalization chips.

Integrated fluidic circuits are constructed from polydimethylsiloxane (PDMS) using multilayer soft lithography (MSL). This technology integrates fluid lines, valves, and reaction chambers into a single chip, enabling the partitioning of the reaction into numerous separate microunits [22]. In contrast, the arrayed lipid bilayer chamber system (ALBiC) feature's reaction microchambers arranged on the chip, allowing the reaction to flow directly through an orifice.

The self-priming compartmentalization chip consists of two glass coverslips. By sliding these coverslips, the reaction mixture can be partitioned into microunits. Droplets separated by cdPCR are consistent in volume, resistant to solvent evaporation, and stable. Two commercial cdPCR systems are QuantStudio3D and BioMark. The BioMark system utilizes integrated fluidic circuits with 10,000–40,000 microchambers, while the QuantStudio3D features an integrated chip with 20,000 microchambers [23].

1.4 Droplet ddPCR

Droplet digital PCR (ddPCR) uses water-in-oil and microfluidic technology to create micro-reaction units. Nucleic acids are arbitrarily distributed into water-in-oil droplets, which are then amplified by PCR separately. A two-color optical detection system monitors the signals from each droplet. The chips used in ddPCR can have shapes such as "+," "T," and "Y" [24].

While ddPCR droplets are created with micrometer precision and are generally less expensive than cdPCR, they are not always uniform in volume. To address this, commercial ddPCR systems use standardized procedures to select droplets with similar volumes for fluorescence detection, and results are adjusted using Poisson statistics. Two prominent ddPCR vendors are QX100/200, which can divide each reaction mixture into 20,000 droplets, and Rain-Drop, which can partition mixtures into 1000,000 to 10,000,000 droplets [25].

2 Materials

1. Droplet digital PCR (Bio-rad QX100TM).
2. Primers and probes.
3. ddPCR supermix.

3 Methods

3.1 Sample Preparation

1. In the very first step, PCR-ready samples are prepared. Nucleic acid samples, primers and probes are combined with propriety supermix (Bio-Rad ddPCR supermix).
2. 20 μL of the prepared samples are loaded into individual wells of the eight-channel disposable droplet generator cartridge.

3.2 Droplet Generation

1. As nucleic acid samples (DNA or RNA) are prepared using primers, fluorescent probes (TaqMan probes with FAM and HEX or VIC), and a ddPCR supermix, developed specifically for droplet generation, the primary step of this technique is performed.
2. Samples are placed into droplet generator (QX200 Droplet generator), which uses propriety reagents and microfluidics to fractionate the samples into 20,000 nanoliter-sized droplets. The droplets are uniform in size and volume.

3.3 PCR Amplification of Droplets

1. Emulsified samples from the cartridge are pipetted and transferred to a 96-well plate for PCR amplification in any compatible thermal cycler.
2. PCR is performed to end point (40 cycles) in a thermal cycler.

3.4 Droplet Reading

1. Following PCR amplification of the nucleic acid target in the droplets, samples are placed in droplet reader (QX200 Droplet reader), which analyzes each droplet individually using a two-color detection system (set to detect FAM and either HEC or VIC).

2. This enables multiplexed analysis for different targets in the sample.
3. The droplet reader along with its software (QuantaSoftTM) counts the PCR-positive and PCR-negative droplets.

3.5 Result Analysis

1. ddPCR software reads the positive and negative droplets in each sample and plots the fluorescence droplet by droplet. Approx. 1.4 million droplets are read per 96-well plate.
2. Positive droplets, containing at least one copy of the target exhibit increased fluorescence over negative droplets.
3. The software (QuantaSoftTM) measures the numbers of droplets that are positive and negative for each fluorophore (FAM and HEX or VIC) in the sample.
4. The fraction of positive droplets is then fitted to a Poisson distribution to determine the absolute initial copy number of the target DNA molecule in the input reaction mixture in units of copies/μL.

3.6 Graphical representation (Fig. 2)

1. Each droplet in the sample is graphed based on its fluorescence intensity and droplet number. Droplets that exceed a specified threshold (indicated by a red line) are classified as positive and given a score of 1. Conversely, droplets below the threshold are deemed negative and assigned a score of 0. This method of counting yields a digital signal used to determine the initial target DNA concentration through statistical analysis of positive and negative droplet counts in the sample.

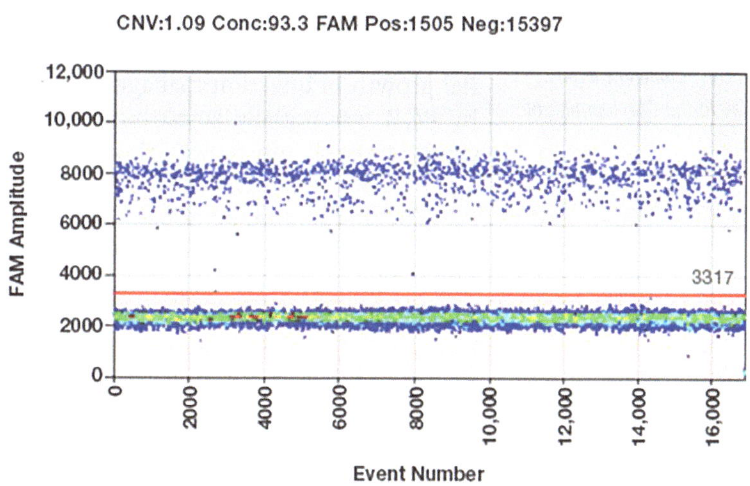

Fig. 2 Graphical representation in ddPCR. (Chap. 26). (Image Courtesy: BIO-RAD/ https:/www.bio-rad.com/en-in/life-science/learning-center/introduction-to-digital-pcr/what-is-droplet-digital-pcr)

4 Notes

1. The reagents and workflows of droplet digital PCR are similar to those used in most standard TaqMan probe-based assays [13].
2. Droplets are created within a water-oil emulsion to generate partitions that isolate the template DNA molecules [41].
3. This technique allows for nanodroplet PCRs that are independent single amplification events [13].
4. ddPCR technology allows for high-throughput digital PCR with reduced sample and reagent volumes, resulting in lower costs compared to alternative methods, all while preserving the sensitivity and precision characteristic of digital PCR [43].
5. ddPCR proves an improved technique over existing digital PCR by means of absolute quantification, unparalleled precision, increased signal-to-noise ratio, removal of PCR bias, simplified quantification, reduced consumable costs, and superior partitioning [41].
6. ddPCR is applied to detect rare DNA target copies with unmatched sensitivity [41].
7. It can be employed to determine the copy number variation with unrivalled accuracy [43].
8. Additionally, it measures gene expression levels with exquisite precision [13].

4.1 Advancements in Digital PCR Technology

4.1.1 Limiting Dilution—Key to the Development of dPCR

Limiting dilution involves diluting a sample to the point where, when distributed across multiple individual assays, the target analyte is observed in some assays but not in others. This technique emerged from studies during World War I for enumerating bacterial growth in infection management. Limiting dilution succeeded by PCR was initially employed to quantify human immunodeficiency proviral molecules. Shortly thereafter, a similar approach was used for quantifying leukemic cells [26].

The first finding demonstrating the overall Pertinence of limiting dilution for quantifying target DNA molecules was issued in 1992. These early investigations used a limited number of standard PCR replicates (fewer than 30) to apply the limiting dilution principle. At that moment, the advancement and commercial deployment of qPCR technology provided a more practical and user-friendly method for PCR-based nucleic acid quantification, leading to a decline in the use of limiting dilution for this purpose [27].

In 1999, a commercially accessible 384-well plate format was used to quantify disease-associated mutations in patients with colorectal cancer by applying limiting dilution to a greater number of replicates.

The term "digital PCR" was introduced by the authors [10]. Despite this innovation, the technique was still considered cumbersome compared to qPCR and did not see widespread adoption at that time. By 2002, advancements in microfluidics engineering enabled the advancement of microfluidic chips that facilitated simple, quick, and scalable fabrication [28]. These advancements made digital PCR a more straightforward and feasible method for precise nucleic acid quantification. As a result, the first articles using commercial microfluidic equipment began to highlight the prospects of dPCR technology [12].

4.2 Commercial dPCR Systems

Fluidigm Corporation used their integrated fluidics technology to launch BiomarkTM, a commercial dPCR platform, in 2006.This platform featured array chips with 12 panels, each containing 765 partitions.A second chip including 48 panels, each panel having 770 partitions was developed by Fluidigm in 2009, resulting in a reduction in expenses and an improvement in throughput. In 2009, Fluidigm introduced a second chip featuring 48 panels, each with 770 partitions, enhancing throughput and lowering technology expenses.

The OpenArray® technology was introduced by Life Technologies in 2009. It consists of tiny plates with through-holes larger than 3000 microns in size that hold the PCR mix in place by surface tension. To streamline processes and cut expenses, they unveiled high-density nanofluidic chips in 2013 that could store up to 20,000 data points.

In 2014, Formulatrix released the Constellation system, which employs an exclusive 96-well microfluidic device. Each well in this system includes 496 chambers for performing individual PCR assays. The introduction of droplet-based dPCR commercial models facilitated the broader adoption of dPCR machinery. The workflow for droplet dPCR comprises creating a water-in-oil emulsion where aqueous droplets contain the PCR mix. Precise adjustments to PCR and oil chemistry allow for the generation of thousands and even millions of uniform nanoliter or picoliter-sized droplets. These droplets are designed to withstand rapid flow through thermal cycling and microfluidic channels without merging or bursting. As of late 2011, Bio-Rad Laboratories introduced the QX100 droplet dPCR system, employing probe-based chemistry for detection. In the following year, Bio-Rad Laboratories released the QX200 droplet dPCR method, which employs both probe-based and intercalating dye chemistries for detection. Both methods make around 20,000 nanoliter-sized droplets from each PCR mixture, providing a more cost-effective solution than previous array chip-based systems.

RainDance Technologies launched the RainDrop® system in 2012, which is a dPCR platform generating millions of picoliter-sized droplets from each PCR mixture.

Different dPCR platforms can vary in terms of workflow, partition volume, the number of partitions, and sample throughput. Additionally, the fraction of the PCR mix that is interpreted by dPCR differs among platforms. Some models retain a ration of the PCR mix within the microfluidics and do not partition it. Moreover, partitions may be omitted from exploration if they do not meet certain attributes. These features can offer specific advantages or disadvantages depending on the purpose. When selecting a dPCR instrument, it is important to consider the vital dynamic range, sensitivity, precision, throughput, and per-sample operating costs for the intended application [29].

5 Core Concepts of Digital PCR

5.1 Methodology

The methodology of dPCR is typically straightforward. In chip-based methods, the PCR mix is typically equally dispersed across preformed partitions, usually with the help of loading equipments. For PCR amplification, the loaded chip is thereafter put in a thermal cycler. After heat cycling, each partition in the chip is imaged using a camera to determine which partitions contain target molecules (positive partitions) and which do not (negative partitions).

A threshold is established to categorize partitions as negative or positive using a normalized fluorescence intensity signal. A Cq value is provided for each partition by certain chip-based models that also gather data in real time.

In droplet-based systems, partitions are created interactively by the user: oil and PCR mix are usually pipetted into wells of a cartridge, which is then placed into specialized instruments that generate a water-in-oil emulsion. The droplet emulsion is then put into a thermal cycler for amplification after being directed into a well in a 96-well plate or a tube in a tube strip. Finally, the tubes or plate are moved to a droplet reading device.

Here, droplets are suctioned and rapidly passed by a fluorescence detector, resembling a flow cytometer, to measure fluorescence in each droplet. Using the fluorescence signal, a threshold is established to the fluorescence signal to categorize the droplets as either negative or positive.

5.2 Estimating Copy Number

Poisson statistics are used to evaluate the average number of molecules per partition by taking into consideration the probability that a given partition includes zero, one, or more molecules [28].

Poisson statistics can also be used to evaluate the concentration of target nucleic acid molecules in the PCR mix if the average partition volume, or VP, is known using Eq. 1.

{Target DNA} = −In(1 − N P/N T) 1/V P copy number per unit volume

where [TargetDNA] represents the concentration of target DNA molecules/unit volume of PCR mix, Positive partition numbers are represented by NPN_PNP, and Partition total is denoted by NTN_TNT.

The precision of the estimation of nucleic acid concentration depends on the accurate measurement of each element in the equation.

5.3 Dynamic Range and Precision

The hypothetical dynamic range of dPCR is constrained by the number of partitions observed. Precision may be harmed at both the lower and upper limits of the dynamic [30] even though the partitioning procedure accompanies binomial distribution, allowing the dynamic range to surpass the number of partitions investigated. The precision and dynamic range of digital PCR (dPCR) might also be impacted by the consistency of partition size.

For optimal precision, monodisperse droplets are necessary. In contrast, techniques that use polydisperse partition sizes tend to have lesser accuracy because larger droplets are more likely to contain multiple targets compared to minor droplets within the same sample. This variability can affect the Poisson statistics calculation, which assumes that all partitions are of equal volume.

Presuming uniform partition sizes, a general dPCR assay with 20,000 partitions offers a hypothetical dynamic range spanning several orders of magnitude, up to around 105 molecules [31]. Some commercially available dPCR instruments can generate millions of partitions, thereby multiplying the dynamic range by more than a scale difference of 106 [32].

Provided there is a sufficient amount of biological sample, a large dynamic range is advantageous for detecting rare target molecules amid a background of nontarget molecules. This is because nontarget molecules are dispersed among several partitions, plummeting the likelihood of interference with the detection of rare target molecules [33].

The theoretical precision of digital PCR (dPCR) can be determined using Poisson statistics and improves as the number of partitions increases, assuming the evaluation is conducted within the optimal window of the dynamic range. The dPCR generally provides higher precision and better resolving power compared to quantitative PCR (qPCR) and other technologies like fluorescence in situ hybridization [34]. This enhanced precision makes dPCR particularly beneficial for uses like measuring copy number variation.

However, the theoretical precision predicted by the Poisson model may not always match the experimental precision perceived from replicate analyses, as the Poisson model is just one factor among many that affect the overall experimental precision.

Generally, if the number of partitions is fewer than 1000, the Poisson model estimate of copy number will have a coefficient of variation of at least 4%, which can be a major source of experimental imprecision. Conversely, if the number of partitions exceeds 10,000, the coefficient of variation can drop below 1.5%, and other aspects may then become the primary foundations of imprecision.

5.4 Partitioning and Inhibitory Effects

Digital PCR (dPCR) is less susceptible to inhibition compared to quantitative PCR (qPCR), when it comes to inhibitory compounds that are frequently present in food [35],clinical (Sedlak et al.), and environmental samples [37]. Inhibitors can affect the precise quantification of target molecules in conventional qPCR by decreasing amplification efficacy and raising the Cq value [38]. In contrast, dPCR can often tolerate moderate levels of inhibitors without compromising accuracy. This is because dPCR accuracy depends on distinguishing between positive and negative partitions after end-point PCR, which remains feasible even if amplification efficiency is somewhat reduced.

The partitioning process in digital PCR (dPCR) significantly reduces the likelihood of inhibitory impacts due to "cross talk," or cross-reactivity, which are common issues in multiplex qPCR assays. However, if inhibitory substances are present in high concentrations, they can still diminish amplification proficiency to the extent that some partitions may be misclassified, leading to an imprecise assessment of the original template concentration. Given that various PCR tests are prone to inhibition varies and the mechanisms of interference can differ among inhibitory compounds [37], it is crucial to empirically evaluate prospective inhibitory effects through well-designed control tests before starting dPCR methods.

5.5 Partitioning and Molecule Segregation

Accurate quantification in digital PCR (dPCR) relies on the random dissemination of target nucleic acid molecules across partitions, as this is the Poisson statistics model's underlying premise. Nonrandom partitioning, such as when target sequences are associated or arranged in tandem, can lead to an underestimation of the number of target molecules, as linked copies may end up in the same partition. To address this issue, restriction digest treatment of samples before partitioning can be employed to distinct linked target sequences [39]. For effective restriction digest treatment, prior knowledge of restriction enzymes that do not cleave the target sequence is crucial.

Additionally, appraisal of the restriction enzyme and buffer's effects on the performance of subsequent dPCR assays is also required. Conversely, overestimating the copy number is possible, if some DNA molecules are single-stranded. During partitioning, single-stranded nucleic acids can lead to up to a two times upsurge in the concentration estimate, as the single-stranded conformation can generate up to two times more independently segregated amplifiable templates [40].

5.6 The Partition Volume Effect

The average number of copies per partition is divided by the partition volume to determine the copy number concentration using dPCR (see Eq. 1). Three partition volume-related factors can affect the accuracy of dPCR measurements: the accuracy of the partition volume specified by the manufacturer, variability in the average partition volume between replicate dPCR runs (inter-reaction variability), and variability in partition volume within a single dPCR run (intra-reaction variability).

Both the present chip-based and droplet-based dPCR systems' partition volumes may vary slightly from the volume specified by the manufacturer and if not corrected, this can introduce bias in copy number concentration estimates [41] .Variability in partition volume between different reactions (inter-reaction variability) and within a single reaction (intra-reaction variability) will impact the accuracy and, to a lesser extent, the precision of dPCR concentration estimates [42].

Variability in partition volume affects the detected technical precision and influences how closely technical precision aligns with the theoretical precision estimated by Poisson statistics.

The effect of partition volume fluctuation is minimized in copy number ratio measurements between two experiments performed in a duplex dPCR format. Nevertheless, especially for clinical applications that require absolute estimation of nucleic acids, partition volume variability remains a critical factor [43].

6 Validation of dPCR Assays

6.1 Validation, Verification, and Measurement Uncertainty Definitions

Verification is defined as "providing objective evidence that a given item meets specified requirements," while validation is defined as "verification, where the specified requirements are suitable for the intended use." Verification verifies that the performance characteristics of the measurement system—such as a dPCR assay that targets a particular gene sequence—are met in the context of digital PCR (dPCR) measurements. Alternatively, validation proves that the dPCR test is suitable for the purpose for which it was designed.

Regulatory permission typically needs a check to be confirmed for a specific resolution using defined reagents, tools, and controls. Revalidating the modified system may be required if any specifications are changed or amended to make sure it is still "fit-for-purpose" and suitable for its intended use.

Furthermore, validated performance features might need to be examined on a regular basis to make sure that the required performance qualities are still met. For instance, while utilizing various batches of reagents. It is essential to identify and measure sources of analytical variation and potential bias during analytical validation. An assessment of measurement uncertainty is the most accurate way to depict the combined impact of various sources.

Measurement uncertainty is well-defined as a "non-negative parameter that characterizes the spread of values ascribed to a measurand, founded on the available data." In simpler terms, measurement uncertainty is defined as "the calculation of value range within which the actual value is likely to fall." The estimation of measurement uncertainty for quantitative DNA quantities from qPCR has been studied [44], and many of these principles are also relevant to dPCR measurements. The next section will briefly cover key aspects to reflect when evaluating the execution of dPCR experiments.

6.2 Considerations for Validation of dPCR Assays

To analytically recognize aspects that may impact dPCR assay performance, it is essential to evaluate each step of the method. The steps involved in a typical dPCR process are sample preparation (such as nucleic acid extraction), PCR mix setup, PCR mix partitioning, target amplification via thermal cycling, partition counting, and classification of partitions into negative and positive populations centered on end-point fluorescence signals, and utilizing the known partition volume and Poisson statistics to estimate the target sequence concentration. The setup of the reaction and sample preparation for dPCR are comparable to those of other quantitative PCR techniques, dPCR and qPCR can both be impacted by a variety of factors that cause variation in these processes. As a means of assessing extraction efficiency for absolute quantification, control materials such as spike-in or positive extraction controls can be employed. However, the selection of the positive control must be done with care, as inhibitors may impact different assays in various ways.

Physical techniques such as shredding [45], and ultrasonication [46] have been successfully applied to fragment DNA for dPCR of genomic DNA, allowing partitioning and effective amplification of target regions. Additionally, using appropriate restriction endonuclease digestion can also be beneficial.

From this stage onward, dPCR differs from previous PCR techniques and introduces distinct causes of variation and bias due to its partitioning procedure and utilizing binary end-point fluorescence signals for quantitative analysis and use of binary end-point fluorescence signal detection for quantification. The nucleic acid sample is pipetted and mixed into a master mix, which normally contains primers, buffer, probe, and enzyme, to prepare the dPCR mix. While some degree of pipetting imprecision is unavoidable, implementing calibrated pipettes can mitigate this problem and greater quantities. Gravimetric dilutions with a calibrated balance can be used for measurements with extremely high accuracy [47].

Sample heterogeneity and, particularly at extremely low template DNA concentrations, stochastic effects can affect the precision of replicates. For dPCR measurements, it is crucial to use probe and primers sequences that are extremely specific to the target nucleic acid sequence to ensure reproducibility and performance. Since recovering dPCR amplicons for confirmatory analysis is often impractical, the specificity of the oligonucleotide primers and probes should be initially verified using qPCR with material control, if accessible, and/or via amplicon nucleotide sequencing generated from qPCR experiments. Accuracy is a key concert attribute for validating any quantitative analytical technique.

The proper partitioning of samples into positive and negative populations is essential for the accurate estimation of nucleic acid molecules in dPCR, as the fraction of positive partitions is a key element in calculating concentration of copy numbers (Eq. 1). A mathematical simulation study assessing variance components in dPCR found that improper partition classification is a major source of inaccuracy in nucleic acid quantification outcomes [48].Therefore, one of the main objectives of dPCR assay optimization is to limit the number of intermediate fluorescence intensity partitions and enhance the difference in fluorescence amplitude between positive and negative partitions. This can be accomplished by improving the dPCR assay's efficiency.

Although dPCR may tolerate less-than-ideal amplification efficiency compared to qPCR, a sensitive and efficient dPCR assay reduces the peril of misclassifying partitions, particularly false negatives. Reference materials with predetermined copy number concentrations should be used to refine the assay, select the optimal cycle number, and adjust the fluorescence threshold accordingly. If suitable reference materials are not available, using two or more assays that span different regions of the target sequence can help verify that ideal settings have been achieved and that the experiment is not affected by DNA secondary structure. Additionally, to maximize the segregation between the fluorescence intensities of the positive and negative divisions, thermal cycling settings should be improved by implementing an annealing temperature gradient. Following the annealing temperature gradient, a denaturation temperature gradient can be applied, especially for templates with high GC content. Probe and primer concentrations should be optimized using methods similar to those used for qPCR. For dPCR assays utilizing intercalating dye chemistry, optimizing primer concentration is also necessary. Conditions optimized for dPCR may not necessarily be optimal for qPCR; thus, if ideal conditions for qPCR have already been determined, they should be confirmed for dPCR as well.

Redesigning the assays can be required in some circumstances. Following optimization, a positive fluorescence signal should be produced by partitions having at least one target molecule, and a negative signal by partitions lacking the target. However, in practice, dPCR assays can produce false negative as well as false positive signals. The threshold setting plays a critical role in determining the rates of false positives and false negatives.

The reliability of assays when dealing with small number of target molecules can be greatly impacted by a small percentage of false positives, while the effectiveness of assays with numerous target molecules can be severely impacted by a small percentage of false negatives.

For validating dPCR readings, the limits of detection (LoD) and blank (LoB) are essential performance metrics. "The lowest amount of analyte that can be detected" is the definition of LoD, while "the maximum measurement value expected for a blank sample" is the definition of LoB, with the corresponding probability indicated [49].

The typical method for determining LoD is an array of measurements of samples with extremely low amounts of analyte, and for LoB, it is determining the distribution of results from many measurements of blank samples. For the blank signal, the LoB is usually found at the 95th percentile of its distribution. Compared to the LoB, the LoD will be greater [43].

In the context of dPCR, LoD can be characterized as the minimum quantity of target nucleic acid molecules that exhibits statistical differentiation from the negative or background control.

A study aimed at establishing lower limits of detection for mutations linked to cancer has experimentally demonstrated one technique for assessing LoD and LoB for a dPCR assay [50].

It is essential to confirm and track instrument performance before using dPCR as a regular diagnostic procedure. Even when the test protocol is correctly performed, instrument-related factors have the potential to induce bias into the results. For instance, repeatable outcomes during amplification depend on precise and reliable temperature settings.

When ideal temperatures are not reached in 96 wells of a thermal cycler block, inaccurate thermal cycling temperatures that arise from insufficient temperature uniformity throughout the wells can be of consequence in either no dPCR amplification or lower amplification efficacy.

7 Clinical Applications of D-PCR

7.1 Virology Applications

Once the technology was developed, early diagnostic applications focused on quantitative virological techniques, leveraging the technology's advantages in quantification. Consequently, methods were established for various pathogens, including certain GB hepatitis virus, human T-lymphotropic virus type 1 (HTLV-1), Epstein-Barr virus, hepatitis B CCC viral DNA, cytomegalovirus (CMV), and SARS-CoV-2 [51]. For hepatitis B, research has demonstrated that dPCR offers a sensitivity for detecting hepatitis B virus genomes in liver tissue that is approximately ten times more than the existing high sensitivity of qPCR [52].

dPCR has been particularly valuable in studying HIV infection, addressing limitations of previous techniques. Research has utilized dPCR to analyze cell-allied HIV DNA. It has been shown that the assessment of episomal 2-LTR circles and total HIV DNA in cells from infected individuals is significantly more precise and trustworthy when done using ddPCR. This method offers significant advantages, including independence from external standards for quantification and reduced sensitivity to mismatches in probe and primer sequences. A more precise knowledge of the cellular reservoir is made possible by the increased precision in measuring proviral DNA [53]. Several studies have shown that dPCR provides higher sensitivity for detecting HIV RNA, leading to prior detection [54].

The dPCR has also been employed to envisage and monitor disease progression by estimating the copy number of the CCL4L gene that encodes the chemokine. This gene encodes a ligand for CCR5 and acts as an infection suppressor [55]. These methods offer a more inclusive opinion of the various HIV reservoirs and are crucial for understanding the infection better and developing new drugs aimed at eliminating it [2].

dPCR also significantly contributes to the detection of integrated viral genomes. For example, it has advanced the identification of human herpesvirus-6 (HHV-6) that is inherited and chromosomally integrated (ICI). When HHV-6 integrates into germ cells, it causes ICI-HHV-6 to be present in all of the progeny from infected people. dPCR enables precise analysis of viral-to-human DNA ratio in these incidents with incredible precision [56]. According to a recent finding, dPCR can quantify Epstein–Barr virus and cytomegalovirus can be quantified with a sensitivity of approximately 100–150 copies/mL in plasma and serum, with extremely low coefficients of deviation and excellent device equivalence [57].

Additionally, a retrospective analysis on human adenovirus reactivation following HSCT (hematopoietic stem cell transplantation) demonstrated that dPCR outperformed qPCR in terms of reproducibility and sensitivity, while providing comparable quantifiability. dPCR detected DNA in 49 (9.0%) of 545 samples from HSCT patients, whereas qPCR detected it in only 11 samples (2.0%) [58]. However, it has not always been shown that these technical improvements translate into better patient management. A 2014 study found that monitoring of CMV viremia in patients receiving hematopoietic stem cell transplants resulted in more suitable and speedier therapy than did dPCR surveillance [36]. From a practical perspective, dPCR offers notable benefits. As mentioned, it is generally more resistant to inhibition as the amplification of each genome occurs separately. This means that even amplifications with subordinate efficacy are sensed and counted as positive. This characteristic can be particularly important for samples that are more susceptible to inhibition, such as fecal samples [59].

This technical premise also clarifies why dPCR is a more consistent method for quantifying viruses that exhibit noteworthy genetic heterogeneity. By amplifying each genome independently, even those with less-than-ideal complementarity to the probes or primers will generate a detectable signal, albeit weaker than that of optimally matched pairs. This signal is recorded and counted as positive. This approach helps mitigate the risk of under-quantification often seen with certain viral infections, as demonstrated with hepatitis and HIV viruses.

In this context, diagnosing human papillomavirus (HPV) infections using qPCR can face limitations due to nucleotide mismatches at primer-binding sites, which may lead to inadequate annealing and under-detection of some HPV types. Additionally, when analyzing anal samples, potential fecal contamination ascends the risk of amplification inhibition. A study has explored whether touchdown dPCR, leveraging genotype-specific probes for targeting the HPV L1 region and broad-spectrum oligonucleotides, can effectively quantify HPV generally and specifically in anal samples [60].

The touchdown dPCR method successfully sensed one copy each of HPV-18 and HPV-16 without cross-reactivity with 27 different high-risk and low-risk HPV strains. When screening self-collected anal specimens, the assay demonstrated high agreement with a commercial multiplex tr-qPCR, accurately identifying samples positive for HPV-16 or HPV-18, as well as those negative for both HPV strains.

Similarly, dPCR has proven reliable in detecting both the SARS-CoV-2 virus and its variants, which may show enhanced antiviral resistance or unusual behavior from a pathogenic or diagnostic perspective [2, 61]. Research has indicated that dPCR can identify between 2.2 and 3.7 copies per reaction, and can raise the positive detection rate in specimens from SARS-CoV-2 infected individuals by 12.0% when compared to RT-qPCR [62].

Additionally, dPCR was able to detect SARS-CoV-2 in 3 of 26 specimens from nearby companions that were negative by RT-qPCR. Furthermore, dPCR has demonstrated the capability to identify subpopulations of viral mutants present at levels as low as 0.1% of the total viral population [2]. Its superior quantification precision has also enhanced the comparability of findings across different laboratories and improved the integrity of international standards, many of which have been established and calculated by this method [36, 63, 64].

7.2 Bacteriology Applications

In bacteriology, many dPCR-based instruments have been designed for detecting and identifying harmful bacteria, with some already approved for clinical use. Nevertheless, a key challenge in applying dPCR to bacteriological diagnostics is leveraging its theoretical benefits over qPCR—such as resistance to inhibitors, greater sensitivity, and the capacity to identify mutants—particularly in regions where qPCR execution can be significantly improved.

dPCR methods have been created for various applications, such as detecting *Helicobacter pylori* in fecal samples as an alternative to techniques involving labeled CO2 in expired air and fecal antigen tests [65]. Additionally, dPCR has been used to identify different species of Borreli and diagnose *Mycobacterium tuberculosis* in both blood and tissue samples [66]. For Lyme disease, traditional serological methods—combining Western blotting and enzyme immunoassay—have long been the standard but have notable limitations, particularly in the precision of enzyme immunoassays. Typical molecular techniques also have reliability issues, often yielding false positives in cerebrospinal fluid (CSF) and blood. A 2017 study exploring the use of dPCR for Lyme disease diagnosis demonstrated promising results [67], successfully detecting bacterial loads as low as ten microbes per sample.

A more current investigation introduced a multiplex dPCR approach with a sensitivity range of 0.2–5 copies of genome-equivalent DNA per microliter, designed for detecting various species within the Bartonella and Borrelia genera [68].

This is particularly relevant for infections like tuberculosis, where the infected region is hard to reach and the DNA released into the bloodstream is too sparse for efficient detection by qPCR. Recent studies have shown that dPCR can significantly improve diagnostic sensitivity for tuberculous meningitis, threefold increase in sensitivity compared to traditional qPCR (57.4% versus 22.1%), while maintaining comparable specificity [69]. A 2017 study demonstrated that dPCR could identify 100% of cases of both extrapulmonary tuberculosis and pulmonary, in contrast to qPCR, which identified only about 50% of these cases [70].

A recent meta-analysis encompassing 14 studies with over 1500 participants presents a more cautious view, indicating comparable sensitivity between qPCR and dPCR for pulmonary tuberculosis. However, it acknowledges that dPCR demonstrates superior sensitivity in detecting extrapulmonary tuberculosis [71]. A recent study compared dPCR, conventional commercial qPCR, and next-generation sequencing (NGS), across various clinical sample types found, that dPCR exhibited the highest sensitivity at 99%, surpassing NGS at 86% and conventional qPCR at 64% [72]. A recent study involving 74 patients with TB, who were diagnosed through acid-fast staining, dPCR, qPCR, and culture, reported that dPCR had a sensitivity of 100%, whereas qPCR had a sensitivity of 82.4%, and acid-fast bacilli (AFB) smear microscopy had a sensitivity of 41.9% [73].

Recent studies highlight the potential of dPCR in quantifying *M. tuberculosis* resistant subpopulations. For instance, a precise dPCR assay designed to measure subpopulations resistant to isoniazid has shown a sensitivity improvement of over 9% compared to conventional drug susceptibility screening [74].

According to recent assessment of bacterial meningitis, a multiplex dPCR assay that was created to identify eight pathogens—including *Salmonella enterica*, *Escherichia coli*, *Haemophilus influenzae, Streptococcus pneumoniae, Staphylococcus aureus, and Streptococcus agalactiae*—demonstrates a very high correlation with traditional culture methods, achieving 94.4% sensitivity and 100% specificity [75]. However, the panel's design may be surprising, as it includes microorganisms not typically associated with human meningitis, and the study does not compare the dPCR results with conventional RT-PCR, which would offer a relevant technological and speed-based alternative.

dPCR can be particularly beneficial in diagnosing conditions where traditional RT-PCR and bacteriological techniques often struggle, such as due to low bacterial loads or difficulties in culturing microorganisms. This includes diseases like endocarditis, pneumonia acquired in a hospital, and osteoarticular infections. In these cases, where accurate quantification of microbes in challenging, samples like bronchoalveolar lavage (BAL) are crucial for diagnosis, dPCR's capabilities can significantly improve diagnostic outcomes.

In the last several years, dPCR has been utilized for the specific diagnosis of various microbes. For instance, a method for diagnosing Veillonella was developed [76], where dPCR demonstrated a sensitivity of 11.3 copies/μL compared to 100 copies/μL for qPCR, although qPCR offered a broader identification series than ddPCR.

A latest finding [77] introduced a duplex qPCR method for detecting *Stenotrophomonas maltophilia and Burkholderia cepacia* right from the blood of the patient, achieving 100% specificity. The corresponding duplex dPCR assay showed excellent repeatability and was capable of detecting as few as 5.35 copies of *B. cepacia* per reaction and 7.67 copies of *S. maltophilia* per reaction. This high sensitivity was consistently observed in both blood bottle samples and simulated blood.

While these techniques offer benefits such as increased sensitivity, resistance to inhibitors, and reliable quantification, their practical value for large-scale clinical diagnosis remains questionable. The primary concern is whether the translation of these methods into routine clinical practice is truly beneficial. Since the microorganisms in question do not typically cause clinical symptoms that would prompt specific diagnostic testing, and are not common etiological entities of common infectious syndromes, their inclusion in numerous diagnostic panels may not be warranted. This mirrors the current situation where these pathogens are often excluded from qPCR diagnostic panels. The true impact of this new technology on improving existing diagnostic methods will be a key feature in determining its adoption.

In a new research evaluating the use of dPCR for diagnosing *Mycoplasma pneumoniae*, it was found that dPCR had a lower limit of detection at 2.9 copies per reaction, compared to 10.8 copies per reaction for qPCR. The qPCR method found 79 positive samples out of 178 clinical samples, while the dPCR test found 80 positive samples and distinguished them.

Notably, one sample that tested negative by qPCR was detected as positive by dPCR, with three copies of bacteria detected per test [78]. In such scenarios, it is essential to precisely assess the specific benefits that dPCR might offer over qPCR, which is already firmly entrenched, with complete multiplex panels, fast processing, high automation, and decreasing costs.

Another study demonstrated promising findings regarding Brucella species detection using dPCR [79]. The study found that the LOD for dPCR was 1.87 copies per reaction with high repeatability. Among 61 patients who tested positive with the serum agglutination test, both qPCR and dPCR had positive detection rates of 75.4% and 88.5%, respectively.

Notably, only 36.3% of putative sero-negative samples were positive by qPCR, whereas 57.6% of samples were positive by dPCR. These findings suggest that dPCR could be a highly valuable diagnostic technique for patients with clinical evidence of brucellosis who test negative serologically.

Similarly, a recent research highlights the capability of dPCR to detect and measure DNA of *Staphylococcus aureus, Escherichia coli, Enterococcus species, and Klebsiella pneumoniae* in blood samples from patients diagnosed with bacteremia. The study reported a LOD ranging from 0.1 to 1 pg/mL [80], allowing for the detection of these microorganisms in the blood within a 3–4 h window, without the need to wait for conventional blood cultures to become positive. This rapid detection represents a significant advancement in the diagnosis and management of bacteremia.

However, clinical data often do not provide sufficient clues to dubious or specific cause of a bacteremia, the obstacle remains to multiplex these dPCR assays. This would enable the simultaneous detection of the main bacterial species responsible for the majority of bacteremia cases. The efficacy of a multiplex dPCR in identifying bacterial infections in the blood of critically sick COVID-19 patients was assessed in a second trial with a similar methodology [81]. The study developed five panels that simultaneously target a range of pathogens, including *Aeruginosa pseudomonas, Aspergillus pyogenes, Acinetobacter baumannii, Escherichia coli, Phytoplanktonus aureus*, coagulase-negative, Streptococcus, Candida, and Enterococcus species *Burkholderia cepacia*, *Stenotrophomonas maltophilia*, and Staphylococcus, *Marcescens serratia*, Citrobacter spp., *Proteus mirabilis, Plasmodium cloacae, Bacteroides fragilis*, and *Morganella morganii* while blaKPC, blaNDM, blaIMP, mecA, Oxa-48, VanM, and VanA are among the seven antimicrobial resistance genes.

The study involved COVID-19 patients confirmed by RT-PCR who were hospitalized in China over a 3-month period. Out of 200 samples examined, 22.5% confirmed positive for bacterial targets via blood culture, whereas 56.5% detected positive using the dPCR experiments. The ddPCR method demonstrated superior performance compared to blood culture in detecting pathogens, identifying mixed infections, and detecting fungi. Among the 113 cases that were positive by ddPCR, 76 were negative by blood culture, and 3 were positive by both blood culture and ddPCR but showed inconsistent bacterial results.

Out of the 79 episodes, 48 (60.8%) met the standards for possible BSI (bloodstream infection), while 13 (16.5%) were classified as probable BSI. The outstanding 18 cases (22.8%) were considered plausible false positives. Additionally, 11 blood culture-positive samples that fell within the expected range were found negative with ddPCR. Blood cultures that were positive for Gram-positive bacteria produced the majority of these false-negative ddPCR findings, implying that the dPCR settings were still not optimized, especially for detecting Gram-positive bacteria.

Similarly, highly sensitive techniques for identifying antimicrobial resistance genes using dPCR are being constructed [82].One such method screens for the vanA and vanB genes in fecal samples. VanA has detection limits of 46.9 digital copies (dcp)/mL, while vanB has detection limits of 60.8 dcp/mL. The experiments demonstrated good linearity across a range of 4.7×10^1 to 3.5×10^5 dcp/mL for vanA and 6.7×10^2 to 6.7×10^5 dcp/mL for vanB. The sensitivity was 100% for both vanA and vanB. However, the positive predictive value was 100% for vanA but only 34.6% for vanB. The lower predictive value for vanB is attributed by the researchers to the presence of vanB DNA-positive anaerobic bacteria in rectal swabs.

In detecting resistance genes, dPCR can be highly useful for identifying carriers of specific resistance genes, which aids in enhancing isolation strategies and halting the emergence of antimicrobial resistance. However, its utility may be more constrained when it comes to assessing antimicrobial resistance in specific infections. This is due to the vast variability in resistance genes and mechanisms, making it challenging to develop sufficient widespread detection panels.

7.3 Applications in Parasitology and Protozoa

Applications of dPCR have been established for diagnosing protozoa with significant epidemiological effect in specific sections, such as Plasmodium spp. [20, 83, 84] and *Trypanosoma cruzi* [85], yielding positive outcomes. It is important to note that accurately measuring the parasitization index in Plasmodium spp. is crucial for determining the right treatment, and dPCR can provide valuable assistance in this regard.

Some reports have also focused on detecting early genes linked with resistance to various antimalarials [86]. Additionally, research has been conducted on diagnosing protozoa like parasites such as *Ascaris lumbricoides* and *Schistosoma japonicum* [87, 88] and Cryptosporidium [89], and however, in these instances, dPCR has not demonstrated clear benefits over traditional qPCR.

Similar to advances in bacteriology, the range of microorganisms diagnosed using dPCR-based techniques has been growing. A recent finding demonstrated that dPCR can detect Demodex spp. in 71.6% of cases suspected of blepharitis caused by this microorganism. However, in this study, dPCR did not show superior efficacy compared to the traditional method of counting mites [90].

7.4 Epidemiology

PCR has demonstrated significant potential in epidemiological research, particularly through WBE (wastewater-based epidemiology), which is used to monitor the circulation of specific enteric viruses and panels of respiratory diseases within communities. Several WBE studies highlight the advantages of dPCR in this context. For instance, a recent 2-year finding in China discovered that SARS-CoV-2 in wastewater is successfully detected by both RT-dPCR and RT-qPCR.

However, RT-dPCR proved more sensitive, being able to detect lower concentrations of SARS-CoV-2 compared to conventional RT-qPCR [91].

A recent finding from Korea also employed dPCR to detect hepatitis A virus, SARS-CoV-2, NDM-type carbapenemases in wastewater and norovirus. The finding demonstrated a positive correlation between the presence of SARS-CoV-2 RNA and the incidence of COVID-19 cases, highlighting the potential of community-scale pathogen surveillance to identify early signs of infection trends. Additionally, the report showcased dPCR's high sensitivity for detecting antibiotic resistance genes, suggesting that this technology could be valuable for monitoring and potentially predicting the spread of these resistance markers [29].

A third WBE study [92], conducted at two wastewater treatment plants in California, utilized dPCR to measure concentrations of norovirus genogroups I and II, human adenovirus group F, rotavirus nucleic acids, and enteroviruses in wastewater solids over a 26-month period ($n = 459$ samples). The study successfully identifies all viral targets, with the lowest concentration found for rotavirus RNA, and the highest concentrations found for norovirus GII nucleic acids and human adenovirus group F. The results demonstrated that dPCR is a sensitive method for detecting enteric viruses in wastewater, making it a reliable tool for WBE programs. This capability can aid in understanding and managing enteric diseases, potentially helping to reduce their transmission.

However, other studies investigating SARS-CoV-2 in wastewater for epidemiological devotions [93] have confirmed that both qPCR and dPCR exhibit excellent sensitivity and specificity. These studies found no significant differences between the methods regarding limits of detection, limits of quantification, or the number of quantifiable and positive samples. Consequently, further research will likely be needed to determine the practical extent to which the theoretical benefits of dPCR, such as improved sensitivity and resistance to inhibitors, are realized in actual applications.

8 Challenges and Future Prospects of dPCR

However, there are some drawbacks to dPCR platforms that could impact how effectively the method is used. The dynamic variety of dPCR is limited by numerous chambers or analytical droplets available for applying the Poisson distribution, which means that proper sample dilution is necessary to ensure accurate results [94]. While dPCR provides precise quantification of pathogenic RNA and DNA, it cannot assess the replicative competence of the pathogen reservoir. This can lead to an overestimation of results, as pathogens lacking replicative capability are also counted. Additionally, dPCR equipment and reagents are costly, limiting its accessibility to only a few laboratories and research institutions.

Controlling costs and expanding applications in pathogen detection remains a major challenge for companies and researchers. Recently, several advancements in dPCR technology have emerged. For instance, the BioRad QX100 platform enables the analysis of 96 samples in a single run. The Raindrop system partitions samples into 107 droplets, enhancing the sensitivity for quantifying rare targets. Following this, Life Technologies introduced the QuantStudio 3D dPCR system, which disperses samples uniformly among

20,000 distinct reaction wells by use of a high-density nanoliter fluidic chip.

This cdPCR platform creates highly stable and uniform small chambers in an immobilized array format. In 2020, Stilla Technologies launched the first six-color fully automatic dPCR device. This platform improved accuracy in genetic analysis and greatly enhanced the capability for multiplex detection of various nucleic acids. It plays a crucial role in detecting pathogenic infections by enabling the analysis of integration sites and clonal expansion. Multiplex dPCR effectively addresses the challenges of labor-intensive and costly integration site analysis by simultaneously targeting both the pathogen and the adjacent human sequences. This approach facilitates monitoring of clonal expansion of specific pathogen sequences through the quantification of double-positive droplets [95].

The proficiency of multiplex dPCR significantly exceeds that of multiplex qPCR with the same number of fluorescent channels. Additionally, multiplex dPCR allows for the differentiation of specific pathogens at the single-cell level [4]. To address false positives, it is essential to enhance the accuracy of data analysis and threshold setting to match the advancements in multiplexing technology. In 2020, the QIAcuity dPCR platform was introduced, which reduces the time from sample preparation to data analysis to under 2 h.

The concept of dPCR is well-established, and current efforts are focused on developing automated platforms for pathogen quantification. Future advancements are expected to include high-throughput devices that are simple, fast, multiplexed, miniaturized, and precise. Researchers are continuously working on new approaches to overcome existing limitations and enhance the utility of dPCR. Similar to qPCR, dPCR may also evolve toward closed and potentially fully integrated sample-to-answer systems to reduce hands-on time and improve simplicity, speed, and portability.

As a result, ongoing advancements in dPCR are anticipated, with future platforms expected to surpass current instruments in various aspects. At the same time, other techniques may complement dPCR in the future. For instance, unbiased metagenomic sequencing is a high-throughput technique that utilizes next-generation DNA sequencing technologies to detect a wide range of pathogens and recognize rare targets [65]. This technique also allows for the detection of individual molecules and offers even greater sensitivity. However, challenges such as nontarget background interference and PCR inhibitors can hinder the quantification of rare pathogens and the preparation of sequencing libraries [65]. To achieve absolute quantification with unbiased metagenomic sequencing, the use of internal calibrators is necessary [66].

In conclusion, dPCR is an advanced PCR technology that provides precise quantification of pathogen sequences. Numerous findings have highlighted its potential for pathogen detection, and recent research has shown significant progress in this area. dPCR offers highly precise measurements of target loads without the need for internal or external calibration. As a result, dPCR is anticipated to be a crucial tool for rapid pathogen detection, offering high specificity sensitivity, and accuracy in the coming years. We foresee that dPCR will greatly enhance the investigation of multifaceted genetic targets and usher in a new epoch of pathogenic diagnostics.

References

1. Kuypers J, Jerome KR (2017) Applications of digital PCR for clinical microbiology. J Clin Microbiol 55:1621–1628. https://doi.org/10.1128/JCM.00211-17
2. Salipante SJ, Jerome KR (2020) Digital PCR – An emerging technology with broad applications in microbiology. Clin Chem 66:117–123. https://doi.org/10.1373/CLINCHEM.2019.304048
3. Majumdar N, Wessel T, Marks J (2015) Digital PCR modeling for maximal sensitivity, dynamic range and measurement precision. PLoS One 10:e0118833. https://doi.org/10.1371/JOURNAL.PONE.0118833
4. Whale AS, Huggett JF, Cowen S, Speirs V, Shaw J, Ellison S, Foy CA, Scott DJ (2012) Comparison of microfluidic digital PCR and conventional quantitative PCR for measuring copy number variation. Nucleic Acids Res 40. https://doi.org/10.1093/NAR/GKS203
5. Streets AM, Huang Y (2014) Microfluidics for biological measurements with single-molecule resolution. Curr Opin Biotechnol 25:69–77. https://doi.org/10.1016/J.COPBIO.2013.08.013
6. Belmonte FR, Martin JL, Frescura K, Damas J, Pereira F, Tarnopolsky MA, Kaufman BA (2016) Digital PCR methods improve detection sensitivity and measurement precision of low abundance mtDNA deletions. Sci Rep 6. https://doi.org/10.1038/SREP25186
7. Barrett AN, Chitty LS (2014) Developing noninvasive diagnosis for single-gene disorders: the role of digital PCR. Methods Mol Biol 1160:215–228. https://doi.org/10.1007/978-1-4939-0733-5_17
8. Marx V (2014) PCR: paths to sensitivity. Nat Methods 11:241–245. https://doi.org/10.1038/NMETH.2849
9. Voegel T, Nelson L (2018) Quantification of agrobacterium vitis from grapevine nursery stock and vineyard soil using droplet digital PCR. Plant Dis 102:2136–2141
10. Corbisier P, Pinheiro L, Mazoua S, Kortekaas AM, Chung PYJ, Gerganova T, Roebben G, Emons H, Emslie K (2015) DNA copy number concentration measured by digital and droplet digital quantitative PCR using certified reference materials. Anal Bioanal Chem 407:1831–1840. https://doi.org/10.1007/S00216-015-8458-Z
11. Vogelstein B, Kinzler KW (1999) Digital PCR. Proc Natl Acad Sci 96:9236–9241. https://doi.org/10.1073/PNAS.96.16.9236
12. Tan LL, Loganathan N, Agarwalla S, Yang C, Yuan W, Zeng J, Wu R, Wang W, Duraiswamy S (2023) Current commercial dPCR platforms: technology and market review. Crit Rev Biotechnol 43:433–464. https://doi.org/10.1080/07388551.2022.2037503
13. Olmedillas-López S, García-Arranz M, García-Olmo D (2017) Current and emerging applications of droplet digital PCR in oncology. Mol Diagn Ther 21(5):493–510. https://doi.org/10.1007/S40291-017-0278-8
14. Milavec M, Dobnik D, Yang L, Zhang D, Gruden K, Žel J (2014) GMO quantification: valuable experience and insights for the future. Anal Bioanal Chem 406:6485–6497. https://doi.org/10.1007/s00216-014-8077-0
15. Debrand E, Lykoudi A, Bradshaw E, Allen SK (2015) A non-invasive droplet digital PCR (ddPCR) assay to detect paternal CFTR mutations in the cell-free fetal DNA (cffDNA) of three pregnancies at risk of cystic. PloS One 10. https://doi.org/10.1371/journal.pone.0142729
16. Pecoraro S, Berben G, Burns M, Corbisier P, de giacorno m (2019) Overview and recommendations for the application of digital PCR. https://publications.jrc.ec.europa.eu/repository/handle/JRC115736. Accessed 10 Sept 2024
17. Bogožalec Košir A, Muller S, Žel J, Milavec M, Mallory AC, Dobnik D (2023) Fast and accurate multiplex identification and quantification of seven genetically modified soybean lines using six-color digital PCR. Foods. https://doi.org/10.3390/foods12224156

18. Whale AS, Huggett JF, Tzonev S (2016) Fundamentals of multiplexing with digital PCR. Biomol Detect Quantif 10:15–23. https://doi.org/10.1016/j.bdq.2016.05.002
19. Kline M, Duewer DL (2017) Evaluating droplet digital polymerase chain reaction for the quantification of human genomic DNA: lifting the traceability fog. Anal Chem 89(8): 4648–4654. https://doi.org/10.1021/acs.analchem.7b00240
20. Dong L, Li W, Xu Q, Gu J, Kang Z, Chen J, Xu X, Zhang X, Zhang X, Jiang H, Guan M (2023) A rapid multiplex assay of human malaria parasites by digital PCR. Clin Chim Acta 539:70–78. https://doi.org/10.1016/J.CCA.2022.12.001
21. Pavšič J, Žel J, Milavec M (2016) Assessment of the real-time PCR and different digital PCR platforms for DNA quantification. Anal Bioanal Chem 408(2016):107–121. https://doi.org/10.1007/s00216-015-9107-2
22. Ramakrishnan R, Qin J, Jones RC, Suzanne Weaver L (2013) Integrated fluidic circuits (IFCs) for digital PCR. Methods Mol Biol 949:423–431. https://doi.org/10.1007/978-1-62703-134-9_27
23. Song Q, Gao Y, Zhu Q, Tian Q, Yu B, Song B, Xu Y, Yuan M, Ma C, Jin W, Zhang T, Mu Y, Jin Q (2015) A nanoliter self-priming compartmentalization chip for point-of-care digital PCR analysis. Biomed Microdevices 17. https://doi.org/10.1007/S10544-015-9970-9
24. Giuffrida MC, Zanoli LM, D'Agata R, Finotti A, Gambari R, Spoto G (2015) Isothermal circular-strand-displacement polymerization of DNA and microRNA in digital microfluidic devices. Anal Bioanal Chem 407: 1533–1543. https://doi.org/10.1007/S00216-014-8405-4
25. Shuga J, Zeng Y, Novak R, Lan Q, Tang X, Rothman N, Vermeulen R, Li L, Hubbard A, Zhang L, Mathies RA, Smith MT (2013) Single molecule quantitation and sequencing of rare translocations using microfluidic nested digital PCR. Nucleic Acids Res 41:e159–e159. https://doi.org/10.1093/NAR/GKT613
26. Heid CA, Stevens J, Livak KJ, Williams PM (1996) Real time quantitative PCR. Genome Res 6:986–994. https://doi.org/10.1101/GR.6.10.986
27. Thorsen T, Maerkl SJ, Quake SR (1979) Microfluidic large-scale integration. Science 298(2002):580–584. https://doi.org/10.1126/SCIENCE.1076996
28. Dube S, Qin J, Ramakrishnan R (2008) Mathematical analysis of copy number variation in a DNA sample using digital PCR on a nanofluidic device. PLoS One 3. https://doi.org/10.1371/JOURNAL.PONE.0002876
29. Tiwari A, Ahmed W, Oikarinen S et al (2022) Application of digital PCR for public health-related water quality monitoring. Sci Total Environ. https://www.sciencedirect.com/science/article/pii/S0048969722027590. Accessed 10 Sept 2024
30. Bhat S, Herrmann J, Armishaw P, Corbisier P, Emslie KR (2009) Single molecule detection in nanofluidic digital array enables accurate measurement of DNA copy number. Anal Bioanal Chem 394:457–467. https://doi.org/10.1007/S00216-009-2729-5
31. Pinheiro LB, Coleman VA, Hindson CM, Herrmann J, Hindson BJ, Bhat S, Emslie KR (2012) Evaluation of a droplet digital polymerase chain reaction format for DNA copy number quantification. Anal Chem 84:1003–1011. https://doi.org/10.1021/AC202578X/SUPPL_FILE/AC202578X_SI_002.PDF
32. Zhong Q, Bhattacharya S, Kotsopoulos S, Olson J, Taly V, Griffiths AD, Link DR, Larson JW (2011) Multiplex digital PCR: breaking the one target per color barrier of quantitative PCR. Lab Chip 11:2167–2174. https://doi.org/10.1039/C1LC20126C
33. Pekin D, Skhiri Y, Baret JC, Le Corre D, Mazutis L, Ben Salem C, Millot F, El Harrak A, Hutchison JB, Larson JW, Link DR, Laurent-Puig P, Griffiths AD, Taly V (2011) Quantitative and sensitive detection of rare mutations using droplet-based microfluidics. Lab Chip 11:2156–2166. https://doi.org/10.1039/C1LC20128J
34. Belgrader P, Tanner SC, Regan JF, Koehler R, Hindson BJ, Brown AS (2013) Droplet digital PCR measurement of HER2 copy number alteration in formalin-fixed paraffin-embedded breast carcinoma tissue. Clin Chem 59:991–994. https://doi.org/10.1373/CLINCHEM.2012.197855
35. Morisset D, Štebih D, Milavec M, Gruden K, Žel J (2013) Quantitative analysis of food and feed samples with droplet digital PCR. PLoS One 8:e62583. https://doi.org/10.1371/JOURNAL.PONE.0062583
36. Sedlak RH, Kuypers J, Jerome KR (2014) A multiplexed droplet digital PCR assay performs better than qPCR on inhibition prone samples. Diagn Microbiol Infect Dis 80:285–286. https://doi.org/10.1016/J.DIAGMICROBIO.2014.09.004
37. Rački N, Dreo T, Gutierrez-Aguirre I, Blejec A, Ravnikar M (2014) Reverse transcriptase

droplet digital PCR shows high resilience to PCR inhibitors from plant, soil and water samples. Plant Methods 10:1–10. https://doi.org/10.1186/S13007-014-0042-6

38. Huggett JF, Novak T, Garson JA, Green C, Morris-Jones SD, Miller RF, Zumla A (2008) Differential susceptibility of PCR reactions to inhibitors: AN important and unrecognised phenomenon. BMC Res Notes 1:1–9. https://doi.org/10.1186/1756-0500-1-70/TABLES/3
39. Hindson BJ, Ness KD, Masquelier DA, Belgrader P, Heredia NJ, Makarewicz AJ, Bright IJ, Lucero MY, Hiddessen AL, Legler TC, Kitano TK, Hodel MR, Petersen JF, Wyatt PW, Steenblock ER, Shah PH, Bousse LJ, Troup CB, Mellen JC, Wittmann DK, Erndt NG, Cauley TH, Koehler RT, So AP, Dube S, Rose KA, Montesclaros L, Wang S, Stumbo DP, Hodges SP, Romine S, Milanovich FP, White HE, Regan JF, Karlin-Neumann GA, Hindson CM, Saxonov S, Colston BW (2011) High-throughput droplet digital PCR system for absolute quantitation of DNA copy number. Anal Chem 83:8604–8610. https://doi.org/10.1021/AC202028G
40. Bhat S, McLaughlin JLH, Emslie KR (2011) Effect of sustained elevated temperature prior to amplification on template copy number estimation using digital polymerase chain reaction. Analyst 136:724–732. https://doi.org/10.1039/C0AN00484G
41. Hunter-Schlichting DVN, Vogel RI, Geller MA, Nelson HH (2024) Quantification of low-level human cytomegalovirus and Epstein-Barr virus DNAemia by digital PCR. J Virol Methods 325:114876. https://doi.org/10.1016/J.JVIROMET.2023.114876
42. Jacobs BKM, Goetghebeur E, Clement L (2014) Impact of variance components on reliability ofl absolute quantification using digital PCR. BMC Bioinformatics 15:1–13. https://doi.org/10.1186/1471-2105-15-283/FIGURES/9
43. Jennings LJ, George D, Czech J, Yu M, Joseph L (2014) Detection and quantification of BCR-ABL1 fusion transcripts by droplet digital PCR. J Mol Diagn 16:174–179. https://doi.org/10.1016/J.JMOLDX.2013.10.007
44. Griffiths KR, Burke DG, Emslie KR (2011) Quantitative polymerase chain reaction: a framework for improving the quality of results and estimating uncertainty of measurement. Anal Methods 3:2201–2211. https://doi.org/10.1039/C1AY05069A
45. Yukl SA, Kaiser P, Kim P, Li P, Wong JK (2014) Advantages of using the QIAshredder instead of restriction digestion to prepare DNA for droplet digital PCR. BioTechniques 56:194–196. https://doi.org/10.2144/000114159
46. George D, Czech J, John B, Yu M, Jennings LJ (2013) Detection and quantification of chimerism by droplet digital PCR. Chimerism 4. https://doi.org/10.4161/CHIM.25400
47. Pinheiro LB, Coleman VA, Hindson CM, Herrmann J, Hindson BJ, Bhat S, Emslie KR (2012) Evaluation of a droplet digital polymerase chain reaction format for DNA copy number quantification. Anal Chem 84:1003–1011. https://doi.org/10.1021/AC202578X
48. Jacobs BKM, Goetghebeur E, Clement L (2014) Impact of variance components on reliability ofl absolute quantification using digital PCR. BMC Bioinformatics 15. https://doi.org/10.1186/1471-2105-15-283
49. Hindson CM, Chevillet JR, Briggs HA, Gallichotte EN, Ruf IK, Hindson BJ, Vessella RL, Tewari M (2013) Absolute quantification by droplet digital PCR versus analog real-time PCR. Nat Methods 10:1003–1005. https://doi.org/10.1038/NMETH.2633
50. Milbury CA, Zhong Q, Lin J, Williams M, Olson J, Link DR, Hutchison B (2014) Determining lower limits of detection of digital PCR assays for cancer-related gene mutations. Biomol Detect Quantif 1:8–22. https://doi.org/10.1016/J.BDQ.2014.08.001
51. Duong K, Ou J, Li Z, Lv Z, Dong H, Hu T, Zhang Y, Hanna A, Gordon S, Crynen G, Head SR, Ordoukhanian P, Wang Y (2021) Increased sensitivity using real-time dPCR for detection of SARS-CoV-2. BioTechniques 70: 7–20. https://doi.org/10.2144/BTN-2020-0133
52. Wang Z, Chen YM, Deng H, Zhen X, Xiong J, Hu Y (2022) Quantification of intrahepatic cccDNA in HBV associated hepatocellular carcinoma by improved ddPCR method. J Virol Methods 299:114334. https://doi.org/10.1016/J.JVIROMET.2021.114334
53. Tumpach C, Cochrane CR, Kim Y, Ong J, Rhodes A, Angelovich TA, Churchill MJ, Lewin SR, Telwatte S, Roche M (2023) Adaptation of the intact proviral DNA assay to a nanowell-based digital PCR platform. J Virus Erad 9:100335. https://doi.org/10.1016/J.JVE.2023.100335
54. Kiselinova M, Pasternak AO, De Spiegelaere W, Vogelaers D, Berkhout B, Vandekerckhove L (2014) Comparison of droplet digital PCR and seminested real-time PCR for quantification of cell-associated HIV-1 RNA. PLoS One 9:e85999. https://doi.org/10.1371/JOURNAL.PONE.0085999

55. Bharuthram A, Paximadis M, Picton ACP, Tiemessen CT (2014) Comparison of a quantitative real-time PCR assay and droplet digital PCR for copy number analysis of the CCL4L genes. Infect Genet Evol 25:28–35. https://doi.org/10.1016/J.MEEGID.2014.03.028
56. Hill JA, HallSedlak R, Magaret A, Huang ML, Zerr DM, Jerome KR, Boeckh M (2016) Efficient identification of inherited chromosomally integrated human herpesvirus 6 using specimen pooling. J Clin Virol 77:71–76. https://doi.org/10.1016/J.JCV.2016.02.016
57. Sedlak RH, Hill JA, Nguyen T, Cho M, Levin G, Cook L, Huang ML, Flamand L, Zerr DM, Boeckh M, Jerome KR (2016) Detection of human herpesvirus 6B (HHV-6B) reactivation in hematopoietic cell transplant recipients with inherited chromosomally integrated HHV-6A by droplet digital PCR. J Clin Microbiol 54:1223–1227. https://doi.org/10.1128/JCM.03275-15/ASSET/B7ADBBC1-E6C3-4A49-9F6A-0C3258A2D450/ASSETS/GRAPHIC/ZJM9990949080002.JPEG
58. Haruta K, Takeuchi S, Yamaguchi M, Horiba K, Suzuki T, Torii Y, Narita A, Muramatsu H, Takahashi Y, Ito Y, Ichi Kawada J (2023) Droplet digital PCR development for adenovirus load monitoring in children after hematopoietic stem cell transplantation. J Mol Diagn 25:403–409. https://doi.org/10.1016/j.jmoldx.2023.03.004
59. Gu J, Ji H, Liu T, Chen C, Zhao S, Cao Y, Wang N, Xiao M, Chen L, Cai H (2022) Detection of cytomegalovirus (CMV) by digital PCR in stool samples for the non-invasive diagnosis of CMV gastroenteritis. Virol J 19:1–10. https://doi.org/10.1186/S12985-022-01913-Z/FIGURES/5
60. Haqshenas G, Garland SM, Balgovind P, Cornall A, Danielewski J, Molano M, Machalek DA, Murray G (2023) Development of a touchdown droplet digital PCR assay for the detection and quantitation of human papillomavirus 16 and 18 from self-collected anal samples. Microbiol Spectr 11. https://doi.org/10.1128/SPECTRUM.01836-23/SUPPL_FILE/SPECTRUM.01836-23-S0006.DOCX
61. Holland SC, Holland LA, Smith MF, Lee MB, Hu JC, Lim ES (2023) Digital PCR discriminates between SARS-CoV-2 omicron variants and immune escape mutations. Microbiol Spectr 11. https://doi.org/10.1128/SPECTRUM.05258-22/SUPPL_FILE/SPECTRUM.05258-22-S0002.XLSX
62. Kim Y, Lee E, Kim B, Cho J, Ryu SW, Lee KA (2024) Evaluation of diagnostic performance of SARS-CoV-2 infection using digital droplet polymerase chain reaction in individuals with or without COVID-19 symptoms. Clin Chim Acta 554:117759. https://doi.org/10.1016/J.CCA.2023.117759
63. Vesper H, Emons H, Gnezda M, Jain CP, Miller WG, Rej R, Schumann G, Tate J, Thienpont L, Vaks JE, (2011) Characterization and qualification of commutable reference materials for laboratory medicine – approved guideline, clinical and laboratory standards institute (CLSI) guidelines – C53A. https://publications.jrc.ec.europa.eu/repository/handle/JRC47925. Accessed 10 Sept 2024
64. Hayden RT, Su Y, Tang L, Zhu H, Gu Z, Glasgow HL, Sam SS, Caliendo AM (2024) Accuracy of quantitative viral secondary standards: a reexamination. J Clin Microbiol 62. https://doi.org/10.1128/JCM.01669-23/SUPPL_FILE/JCM.01669-23-S0001.DOCX
65. Talarico S, Safaeian M, Gonzalez P, Hildesheim A, Herrero R, Porras C, Cortes B, Larson A, Fang FC, Salama NR (2016) Quantitative detection and genotyping of helicobacter pylori from stool using droplet digital PCR reveals variation in bacterial loads that correlates with cagA virulence gene carriage. Helicobacter 21:325–333. https://doi.org/10.1111/HEL.12289
66. Cresswell F, Lange C, van Crevel R (2020) Improving the diagnosis of tuberculous meningitis: good, but not good enough. Clin Microbiol Infect 26:134–136. https://doi.org/10.1016/j.cmi.2019.10.009
67. King JL, Smith AD, Mitchell EA, Allen MS (2017) Validation of droplet digital PCR for the detection and absolute quantification of Borrelia DNA in Ixodes scapularis ticks. Parasitology 144:359–367. https://doi.org/10.1017/S0031182016001864
68. Maggi R, Breitschwerdt EB, Qurollo B, Miller JC (2021) Development of a multiplex droplet digital PCR assay for the detection of babesia, bartonella, and borrelia species. Pathogens 10:1462. https://doi.org/10.3390/PATHOGENS10111462
69. Li Z, Pan L, Lyu L, Li J, Jia H, Du B, Sun Q, Zhang Z (2020) Diagnostic accuracy of droplet digital PCR analysis of cerebrospinal fluid for tuberculous meningitis in adult patients. Clin Microbiol Infect 26:213–219. https://doi.org/10.1016/j.cmi.2019.07.015
70. Yang J, Han X, Liu A, Bai X, Xu C, Bao F, Feng S, Tao L, Ma M, Peng Y (2017) Use of digital droplet PCR to detect mycobacterium tuberculosis DNA in whole blood-derived DNA samples from patients with pulmonary and extrapulmonary tuberculosis. Front Cell

Infect Microbiol 7:227403. https://doi.org/10.3389/FCIMB.2017.00369/BIBTEX

71. Meregildo-Rodriguez ED, Asmat-Rubio MG, Vásquez-Tirado GA (2023) Droplet digital PCR vs. quantitative real time-PCR for diagnosis of pulmonary and extrapulmonary tuberculosis: systematic review and meta-analysis. Front Med (Lausanne) 10:1248842. https://doi.org/10.3389/FMED.2023.1248842/BIBTEX
72. Zhang D, Yu F, Han D, Chen W, Yuan L, Xie M, Zheng J, Wang J, Lou B, Zheng S, Chen Y (2023) ddPCR provides a sensitive test compared with GeneXpert MTB/RIF and mNGS for suspected mycobacterium tuberculosis infection. Front Cell Infect Microbiol 13:1216339. https://doi.org/10.3389/FCIMB.2023.1216339/BIBTEX
73. Aung YW, Faksri K, Sangka A, Tomanakan K, Namwat W (2023) Detection of mycobacterium tuberculosis complex in sputum samples using droplet digital PCR targeting mpt64. Pathogens 12:345. https://doi.org/10.3390/PATHOGENS12020345
74. Zhang S, Chen X, Lin Z, Tan Y, Liang B, Pan Y, Huang M, Su B, Hu X, Xu Y, Li Q (2023) Quantification of isoniazid-Heteroresistant mycobacterium tuberculosis using droplet digital PCR. J Clin Microbiol 61:e0188422. https://doi.org/10.1128/JCM.01884-22/SUPPL_FILE/JCM.01884-22-S0001.PDF
75. Yang Y, Hu Y, Yang C, Shi W, Jin S, Hua C, Jiang K (2024) Development and validation of a novel multiplex digital PCR assay for identification of pathogens in cerebrospinal fluid of children with bacterial meningitis. Clin Chim Acta 554:117787. https://doi.org/10.1016/J.CCA.2024.117787
76. Ding Z, Cui J, Zhang Q, Feng J, Du B, Xue G, Yan C, Gan L, Fan Z, Feng Y, Zhao H, Xu Z, Yu Z, Fu T, Zhang R, Cui X, Tian Z, Chen J, Chen Y, Li Z, Zhong X, Lin Y, Yuan J (2024) Detecting and quantifying Veillonella by real-time quantitative PCR and droplet digital PCR. Appl Microbiol Biotechnol 108:1–13. https://doi.org/10.1007/S00253-023-12861-1/FIGURES/6
77. Liu C, He Z, Kong M, Jin D (2024) Development of a duplex droplet digital PCR assay for the detection of Burkholderia cepacia complex and Stenotrophomonas maltophilia in bloodstream infections. Microbiol Spectr 12. https://doi.org/10.1128/SPECTRUM.03569-23/SUPPL_FILE/SPECTRUM.03569-23-S0001.PDF
78. Zhao H, Yan C, Feng Y, Du B, Feng J, Cui X, Cui J, Gan L, Fan Z, Xu Z, Fu T, Yu Z, Yuan J, Xue G (2023) Absolute quantification of mycoplasma pneumoniae in infected patients by droplet digital PCR to track disease severity and treatment efficacy. Front Microbiol 14:1177273. https://doi.org/10.3389/FMICB.2023.1177273/BIBTEX
79. Liu J, Song Z, Ta N, Tian G, Yang X, Zhao H, Piao D, Fan Y, Zhang Y, Jiang H (2023) Development and evaluation of a droplet digital PCR assay to detect Brucella in human whole blood. PLoS Negl Trop Dis 17. https://doi.org/10.1371/JOURNAL.PNTD.0011367
80. Tedim AP, Merino I, Ortega A, Domínguez-Gil M, Eiros JM, Bermejo-Martín JF (2024) Quantification of bacterial DNA in blood using droplet digital PCR: a pilot study. Diagn Microbiol Infect Dis 108:116075. https://doi.org/10.1016/J.DIAGMICROBIO.2023.116075
81. Li Y, Huang K, Yin J, Tan ZX, Zhou M, Dai J, Yi B (2024) Clinical evaluation of a multiplex droplet digital PCR for pathogen detection in critically ill COVID-19 patients with bloodstream infections. Infection 52:1027–1039. https://doi.org/10.1007/S15010-023-02157-X/TABLES/5
82. Giersch K, Tanida K, Both A, Nörz D, Heim D, Rohde H, Aepfelbacher M, Lütgehetmann M (2024) Adaptation and validation of a quantitative vanA/vanB DNA screening assay on a high-throughput PCR system. Sci Rep 14(1):1–9. https://doi.org/10.1038/s41598-024-54037-5
83. Costa GL, Alvarenga DAM, Aguiar ACC, Louzada J, Pereira DB, de Oliveira TF, Fonseca Júnior AA, Carvalho LH, Ferreira Alves de Brito C, Nóbrega de Sousa T (2022) Improving the molecular diagnosis of malaria: droplet digital PCR-based method using saliva as a DNA source. Front Microbiol 13:882530. https://doi.org/10.3389/FMICB.2022.882530/BIBTEX
84. Mahendran P, Liew JWK, Amir A, Ching XT, Lau YL (2020) Droplet digital polymerase chain reaction (ddPCR) for the detection of plasmodium knowlesi and plasmodium vivax. Malar J 19:1–10. https://doi.org/10.1186/S12936-020-03314-5/FIGURES/6
85. Ramírez JD, Herrera G, Hernández C, Cruz-Saavedra L, Muñoz M, Flórez C, Butcher R (2018) Evaluation of the analytical and diagnostic performance of a digital droplet polymerase chain reaction (ddPCR) assay to detect Trypanosoma cruzi DNA in blood samples. PLoS Negl Trop Dis 12:e0007063. https://doi.org/10.1371/JOURNAL.PNTD.0007063

86. Srisutham S, Suwannasin K, Sugaram R, Dondorp AM, Imwong M (2021) Measurement of gene amplifications related to drug resistance in plasmodium falciparum using droplet digital PCR. Malar J 20:1–12. https://doi.org/10.1186/S12936-021-03659-5/FIGURES/6
87. Pomari E, Piubelli C, Perandin F, Bisoffi Z (2019) Digital PCR: a new technology for diagnosis of parasitic infections. Clin Microbiol Infect 25:1510–1516. https://doi.org/10.1016/j.cmi.2019.06.009
88. Acosta Soto L, Santísima-Trinidad AB, Bornay-Llinares FJ, Martín González M, Pascual Valero JA, Ros Muñoz M (2017) Quantitative PCR and digital PCR for detection of Ascaris lumbricoides eggs in reclaimed water. Biomed Res Int 2017:7515409. https://doi.org/10.1155/2017/7515409
89. Yang R, Paparini A, Monis P, Ryan U (2014) Comparison of next-generation droplet digital PCR (ddPCR) with quantitative PCR (qPCR) for enumeration of cryptosporidium oocysts in faecal samples. Int J Parasitol 44:1105–1113. https://doi.org/10.1016/J.IJPARA.2014.08.004
90. An N, Dou X, Yin N, Lu H, Zheng J, Liu X, Yang H, Zhu X, Xiao X (2024) The use of digital PCR for the diagnosis of Demodex blepharitis. Curr Eye Res 49:33–38. https://doi.org/10.1080/02713683.2023.2265083
91. Chai X, Liu S, Liu C, Bai J, Meng J, Tian H, Han X, Han G, Xu X, Li Q (2024) Surveillance of SARS-CoV-2 in wastewater by quantitative PCR and digital PCR: a case study in Shijiazhuang city, Hebei province, China. Emerg Microbes Infect 13. https://doi.org/10.1080/22221751.2024.2324502
92. Boehm AB, Shelden B, Duong D, Banaei N, White BJ, Wolfe MK (2024) A retrospective longitudinal study of adenovirus group F, norovirus GI and GII, rotavirus, and enterovirus nucleic acids in wastewater solids at two wastewater treatment plants: solid-liquid partitioning and relation to clinical testing data. MSphere 9. https://doi.org/10.1128/MSPHERE.00736-23/SUPPL_FILE/MSPHERE.00736-23-S0001.PDF
93. de la Cruz Barron M, Kneis D, Geissler M, Dumke R, Dalpke A, Berendonk TU (2023) Evaluating the sensitivity of droplet digital PCR for the quantification of SARS-CoV-2 in wastewater. Front Public Health 11:1271594. https://doi.org/10.3389/FPUBH.2023.1271594/BIBTEX
94. Malla B, Shrestha S, Haramoto E (2024) Optimization of the 5-plex digital PCR workflow for simultaneous monitoring of SARS-CoV-2 and other pathogenic viruses in wastewater. Sci Total Environ 913:169746. https://doi.org/10.1016/J.SCITOTENV.2023.169746
95. Wan L, Li M, Law MK, Mak PI, Martins RP, Jia Y (2023) Sub-5-Minute Ultrafast PCR using Digital Microfluidics. Biosens Bioelectron 242:115711. https://doi.org/10.1016/J.BIOS.2023.115711

Chapter 27

Immunocapture Polymerase Chain Reaction (IC-PCR) in Detection of Plant Viruses

Insha Mehraj, Sumiah Wani, Nulevino Iralu, Sahar Saleem, and Aflaq Hamid

Abstract

Of the modified nucleic acid-based detection methods, a newer technique is developed by the combination of serological assay with molecular assay via ELISA and PCR methods. It provides two-way specificity and double sensitivity, thus making it a robust detection method. In the first part, ELISA is performed using specific antibodies and the captured antigen is then subjected to PCR detection using specific primers in the second part of the protocol. This doesn't require the extraction of viral RNA unlike conventional PCR, thus referred as Immuno-PCR or Immunocapture PCR. This method stands well on 3 "S"s of the efficiency of a diagnostic method, i.e., sensitivity, specificity, and speed. However, the cost of the setup is still high, hindering its use in day-to-day laboratory diagnostics.

Key words Serology, ELISA, Immunocapture PCR, Sensitivity, Specificity

1 Introduction

Immunocapture polymerase chain reaction (IC-PCR) is a modified nucleic acid method in combination with serological (enzyme-linked immunosorbent assay) assay to provide two-way specificity in plant virus detection. In IC-PCR methods, virus-specific antibodies are employed that enable the captivity of viral agents in PCR tubes or microtiter plates, followed by PCR or RT-PCR of the captured viral particles. This approach was first reported in animal pathology for hepatitis A virus detection and later successfully used for other RNA virus families [1, 2]. Owing to its nature, this method offers two-way specificity through combined molecular and serological procedures and is convenient for detecting viruses directly from crude plant extracts, eliminating the need for purified or partially purified viruses and nucleic acid isolation [3]. This hybrid method, providing double sensitivity, proves versatile and robust for successful plant virus detection, particularly in plant

Aflaq Hamid et al. (eds.), *Detection of Plant Viruses: Advanced Techniques*, Springer Protocols Handbooks,
https://doi.org/10.1007/978-1-0716-4390-7_27,

species or tissues containing inhibitory substances [4]. For instance, Grapevine fanleaf virus (GFLV), a Nepovirus genus member, was detected using IC-RT-PCR with a recombinant antibody to the GFLV coat protein (CP) [5]. Despite the robust nature of IC-PCR methods, their direct application in field conditions is hindered by the necessity for thermal cyclers.

2 Materials

1. Virus-specific capture antibody.
2. ELISA buffers, viz., wash buffer, blocking buffer, extraction buffer.
3. Detection antibody.
4. PCR setup (components).

2.1 ELISA Procedure

1. Virus specific capture antibody is coated on microtubes or microtiter plate, whichever is used in the procedure.
2. The coating of capture antibody is followed by a round of addition of wash and blocking buffer of the ELISA kit.
3. **Steps 1–2** are followed by the addition of crude sample suspected to be viral infected.
4. This is followed by another step of washing using wash buffer.
5. Following washing step, a secondary detection antibody is used, trapping the viral antigen in between (double antibody sandwich ELISA).
6. This is followed by a third round of washing to eliminate any unwanted binding.
7. The acquired product is subjected to amplification using polymerase chain reaction (PCR) or reverse transcriptase PCR (RT-PCR) procedure or real-time quantitative PCR.

2.2 Polymerase Chain Reaction Procedure

1. Viral antigen captured in crude form in leaf samples through ELISA eliminates the need for RNA extraction unlike in conventional PCR methods.
2. The captured virus (product from ELISA procedure, as discussed in Subheading 2.1) is processed following the general protocol of PCR.
3. cDNA synthesis.
4. PCR/RT-PCR/ qPCR.

3 Methods

3.1 Enzyme-Linked Immunosorbent Assay (ELISA)

1. *Plate Coating:*
   - Virus-specific antibodies are coated on microtubes or microtiter plate.
   - Antibody binding occurs via nonspecific adsorption. It is typically performed overnight at 4 °C.
   - Coating of capture antibody is achieved with a help of a coating buffer (carbonate, bicarbonate, or borate buffer at pH > 9. Before the procedure is initiated, optimal concentration of the coating antibody should be determined.
   - (Note: Starting range of 0.2–20 μg/mL).
2. *Plate Washing*:
   - Wash steps are essential for the removal of unbound reagents.
   - Washing is accomplished using wash buffers. Typical wash buffers include phosphate buffered saline (PBS) and tris buffered saline (TBS).
   - Wash buffer is generally used in a low concentration (0.01–0.1% v/v).
   - The number and duration of washes should be optimized.
3. *Blocking:*
   - Blocking is critical to prevent nonspecific binding. It helps to maximize Signal: Background ratio (S: B).
   - The blocking buffer is typically a weak solution (1–5%) and commonly used buffers are: bovine serum albumin (BSA), non-fat dried milk, gelatine, casein, whole serum, etc.
   - An overnight blocking step is recommended to improve S:B.
4. *Sample Incubation:*
   - Short incubation times are often sufficient for antibody: antigen binding.
   - Incubation temperature should be optimized.
5. *Dilution of test samples:*
   - Dilution of test samples can reduce background effects caused by the nature of the sample material.
6. *Choice of Antibodies:*
   - It requires highly specific antibodies.
   - Capture and detection antibodies should each target a different epitope of the antigen.

7. *Plate Layout and Results:*
   - Positive Controls: Spiked with known concentrations of antigen.
   - Negative Controls: Sample without antigen.
   - All samples should be analyzed in duplicate.
   - Controls and standards should be diluted in the same matrix as the test samples.

3.2 Polymerase Chain Reaction (PCR)

In the second part of this procedure, reverse transcription polymerase chain reaction is carried out conventionally in the given below steps:

- *cDNA synthesis*: The captured antigen (through ELISA) is used to synthesize cDNA using oligo dT primer as:

Reaction component	Stock solution (Concentration)	Final volume (12 μL)
RNA	0.1 ng-5 μg	2.0
Oligo dT primer	–	1.0
Water	–	5.0
Total reaction volume	–	12.0

(Standardization is done as per laboratory conditions, if required)

The reaction mixture is incubated at 65 °C for 5 min in thermal cycler, followed by the reaction as under:

Reaction component	Stock solution (Concentration)	Final volume (20 μl)
Reaction buffer	5×	4.0
DNTPs	10 mM	1.0
Reverse transcriptase enzyme	200 U/μL	2.0
RNase inhibitor (RiboLock)	20 U/μL	1.0
Water	–	12.0
Total reaction mixture	–	20.0

Incubation is done at 42 °C for 60 min and the reaction will be terminated at 72 °C for 5 min.

– *RT-PCR*: RT- PCR is carried out using virus (coat protein)-specific primers for identification of viruses using standard reaction mixture given below:

Reaction component	Stock solution (Concentration)	Final volume (25 μL)
Buffer	10×	2.5
$MgCl_2$	25 mM	1.5
dNTP mix	10 mM	0.5
Taq polymerase enzyme	5 U/μL	0.25
Primer (forward and reverse)	100 mM	0.75(each)
Water	–	18.75

4 Notes

1. Immuno-PCR combines the specificity of ELISA with the sensitivity of PCR.
2. Immunocapture PCR is adaptable to detection of any protein.
3. It provides exponential signal amplification.
4. It provides double sensitivity and has extremely low limit of detection (pg-fg).
5. It is suitable for small sample volumes.
6. It is compatible with complex samples, especially those carrying inhibitory substances.
7. It has fewer incubation steps than ELISA with improved assay reproducibility.
8. The results are obtained rapidly compared to a conventional PCR.
9. It is amenable to multiplexing.
10. Despite the numerous advantages of this technique, the scope is yet limited due to its dependency on conjugation of an antibody to an oligonucleotides.

References

1. Wetzel T, Candresse T, Macquaire G, Ravelonandro M, Dunez J (1992) A highly sensitive immunocapture polymerase chain reaction method for plum pox potyvirus detection. J Virol Methods 39:27–37
2. Nolasco G, de Blas C, Torres V, Ponz F (1993) A method combining immunocapture and PCR amplification in microtiter plate for the detection of plant viruses and subviral pathogens. J Virol Methods 45:201–218

3. Bhardwaj A, Kulshrestha S (2021) Advance methods for the isolation and characterization of plant viruses infecting crops. Viruses
4. Mulholland V (2005) Immunocapture- polymerase chain reaction. In: Burns R (ed) Immunological protocols. Methods in molecular biology™, vol 295. Humana Press. https://doi.org/10.1385/1-59259-873-0:281
5. Koolivand D, Sokhandan-Bashir N, Behjatnia SAA, Jafari Joozani RA (2014) Detection of grapevine fan-leaf virus by immunocapture reverse transcription-polymerase chain reaction (IC-RT-PCR) with recombinant antibody. Arch Phytopathol Plant Protect 47:2070–2077
6. Emma Easthope (2018) A guide to Immuno-PCR. Innova Biosciences
7. Kokko HI, Kivineva M, Karenlampi SO (1996) Singlestep immunocapture RT-PCR in the detection of raspberry bushy dwarf virus. BioTechniques 20:842–846
8. Sefc KM, Leonhardt W, Steinkellner H (2000) Partial sequence identification of grapevine-leafroll-associated virus-1 and development of a highly sensitive IC-RT-PCR detection method. J Virol Methods 86:101–106
9. Helguera PR, Taborda R, Decampo DM, Ducasse DA (2001) Immunocapture reverse transcription-polymerase chain reaction combined with nested PCR greatly increases the detection of Prunus necrotic ring spot in the peach. J Virol Methods 95:93–91

Chapter 28

Reverse Transcription Loop-Mediated Isothermal Amplification (RT-LAMP) for Plant Viral Detection

Sumiah Wani, Shahjahan Rashid, Nulevino Iralu, Parvaiz Ullah, Sajad Un Nabi, and Aflaq Hamid

Abstract

Reverse transcription loop-mediated isothermal amplification (RT-LAMP) has emerged as a powerful molecular diagnostic tool for the rapid and sensitive detection of RNA targets. Due to its speed, simplicity, usability, and affordable equipment needs, LAMP is a technique for amplifying specific DNA sequences. Its main objective is to detect the plant viruses by achieving DNA amplification in an isothermal environment. Loop-mediated isothermal amplification is a single-tube technique that amplifies nucleic acid with high specificity, efficiency, and rapidity under isothermal conditions with a set of six specially designed primers that recognize six to eight distinct sequences of the target DNA. These primers consist of two internal primers (FIP, BIP), two external primers (F3, B3), and two loop primers. In RT-LAMP, RNA is reverse-transcribed into complementary DNA (cDNA), which serves as the template for amplification. The reaction is carried out at a constant temperature, typically around 60–65 °C. RT-LAMP results can be visualized by different ways such as turbidity and colorimetric, or fluorescent indicators can be added to the reaction mixture, enabling real-time monitoring of amplification progress. Turbidity-based detection relies on the visual observation of turbidity or cloudiness in the reaction mixture due to the accumulation of magnesium pyrophosphate, a by-product of the amplification reaction. Positive reactions appear turbid, while negative reactions remain clear. Colorimetric detection involves the addition of dyes such as hydroxynaphthol blue (HNB) or phenol red to the reaction mixture. Positive reactions result in a color change, typically from pink to yellow or orange, due to pH changes resulting from DNA amplification. Fluorescent dyes like SYBR Green I can be used to monitor amplification in real time using a fluorescence detector. Positive reactions exhibit an increase in fluorescence intensity proportional to DNA amplification.

Key words PCR, RT-LAMP, Viruses, SYBR Green

1 Introduction

Reverse transcription loop-mediated isothermal amplification (RT-LAMP) is a one-step nucleic acid amplification method to multiply specific sequences of RNA as well as DNA. LAMP is a novel approach to nucleic acid amplification, which involves a constant temperature throughout the reaction incubation,

Aflaq Hamid et al. (eds.), *Detection of Plant Viruses: Advanced Techniques*, Springer Protocols Handbooks,
https://doi.org/10.1007/978-1-0716-4390-7_28,

eliminating the need for a thermal cycler. The technique's high sensitivity and rapid results make it valuable for various applications, particularly in diagnostics and research. Its ability to detect minute amounts of target DNA or RNA, sometimes as low as a few copies, within a short time frame (usually 5–10 min), contributes to its popularity, which can be executed using minimal resources such as a water bath for incubation, and enhances its practicality and accessibility. Additionally, detection methods involving UV light and agarose gel electrophoresis allow for straightforward visualization of results without requiring specialized equipment. The loop-mediated isothermal amplification (LAMP) reaction utilizes two sets of primers, each specific for a different target region of the nucleic acid being amplified. These primers were helping to increase the specificity of the amplification and can target a specific pathogen or gene of interest. The LAMP reaction needs a DNA polymerase, which helps in band displacement activity, allowing it to perform the reaction at a constant temperature between 60 °C and 65 °C. The initial version of the LAMP was established by Notomi et al. and it utilized a set of six specific primers, including two internal, two external, and two loop-specific primers [1]. This design helped to increase the specificity and accuracy of the amplification reaction and paved the way for the development of LAMP as a powerful tool for nucleic acid amplification and detection. The RT-LAMP has many advantages over other testing methods for pathogen detection such as faster reaction, low time consuming, high accuracy, and easy to perform. The reaction products can be analyzed during the assay in thermal blocks, a water bath, or an incubator and are indicated by an increase in turbidity (real-time monitoring) or a color change, signifying a positive reaction [1, 2].

2 Materials

1. Symptomatic and asymptomatic leaf samples collected from field and subjected to RNA isolation (Stored at −80 °C).
2. Bst polymerase.
3. 10× Isothermal Amplification Buffer.
4. dNTP mix. Use 25 mM stock.
5. $MgSO_4$.
6. LAMP primers, two set of primers.
7. Warm Start LAMP Master mix.
8. LAMP Fluorescent Dye, SYBR Green, EtBr.
9. Heat block or water bath (65 °C).

3 Methods

1. Isolate total DNA (if testing for DNA viruses) or RNA (if testing RNA viruses) from infected plants.
2. Isolate total DNA or RNA from corresponding virus-free plants to serve as negative control.
3. For lamp, first we design two sets of LAMP primers, i.e., FIP, BIP, F3, B3, and two loop primers LF, LB.
4. The LAMP reaction mixture (25 uL) contains 1 uL (about 60 ng) of the template nucleic acids, two thermopol buffers, 1.4 mM each of dNTPs, 0–14 mM $MgSo_4$ (optimum concentration need to be worked out for each set of primers and target DNA) and 0.4–2.4 M betaine (Optimum concentration need to be worked out of each set of primers and target DNA), 200 nM each of the loop primers (B-loop and F-loop), 1 mM $MnCl_2$ calcein, and 8 U of *Bst* DNA polymerase (see Table 1 below).
5. Incubate the tube at 65 °C for 60 min (temperature and time can be optimized for each target and primer combination by performing LAMP/RT-LAMP for different temperatures starting from 58 °C to 60 °C and time from 30 min to 75 min) followed by 80 °C for 5 min to inactivate the *Bst* polymerase.
6. For the optimization of RT-LAMP, the reaction of LAMP was evaluated with different concentrations of MgSo4 (0, 2, 4, 6, 8,10, and 12 mM), betaine(0, 0.2, 0.4, 0.6, 0.8, 1.0, 1.2, 1.4, and 1.6 M), and dNTPs (0, 0.2, 0.6, 1.0, 1.4, 1.8, 2.2, and 2.6 mM), reaction temperature (56, 58, 60, 62, 64, and 66 °C), and reaction time (10, 20, 30, 40, 50, 60, 70, and 80), respectively. The reaction can be stopped by placing the tube at 80 °C for 10 min to inactivate the enzyme [3].

Table 1
Standard protocol for RT LAMP

Component	Volume (25 μl reaction)
10× isothermal buffer	2.5 μL
dNTPs	3.5 μL (1.4 mM final)
$MgSO_4$	1.5 μL (6 mM + 2 mM in buffer=8 mM final)
10× Primers	2.5 μL (1.6 μM FIP/BIP, 0.2 μM F3/B3, 0.4 μM LoopF/B)
Bst 2.0 (8000 U/mL)	1 μL (0.32 U/μL)
Sample	e.g. 1 μL
H_2O	11.5 μL
RNA targets: WS RTx	2 uL

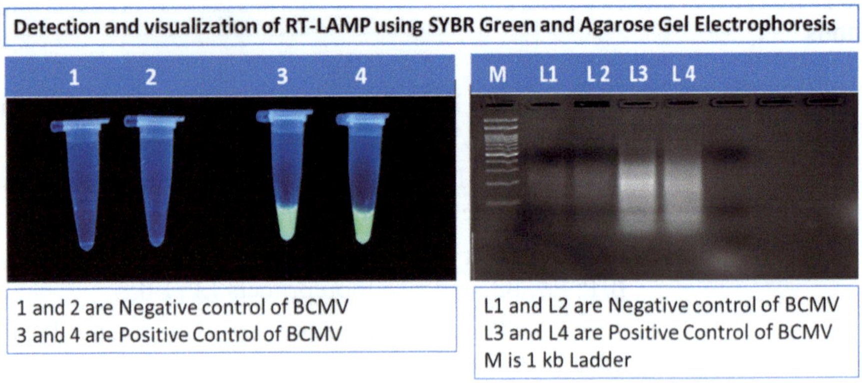

Fig. 1 RT-LAMP detection using SYBR green and agarose gel electrophoresis. (Chap. 28)

3.1 Protocols (Table 1)

3.2 Detection Methods (Fig. 1)

3.2.1 Visual Method

(a) The turbidity of magnesium pyrophosphate, a by-product of the amplification reaction, is produced in proportion to the amount of amplified products. Since LAMP amplification can produce extremely large amounts of amplified products, white turbidity can be visually observed. From this feature, the presence of turbidity can indicate the presence of target gene, and visual detection can be achieved. In LAMP results, the tubes must be observed after the reaction with the naked eye to check the presence of turbidity [4, 5].

(b) If the tube containing the amplified products in the presence of fluorescent intercalating dye (ethidium bromide, SYBER Green, etc.) is illuminated with a UV lamp, the fluorescence intensity increases by [6, 7].

3.2.2 Detection by Electrophoresis

(a) LAMP products are run on a 2% agarose gel. Electrophoresis pattern of LAMP amplified products is not a single band but a ladder pattern because LAMP method can form amplified products of various sizes consisting of alternately inverted repeat of the all-target sequences on the same strand [2].

(b) Real-time detection is also possible by running the LAMP reaction on a real-time LAMP instrument and measuring the fluorescence emitted by SYBR Green or other dyes on a real-time basis). LAMP has been successfully used to detect many infectious agents of humans, animals, and plants. This technique has also been used to detect numerous plant viruses [8].

4 Notes

1. The quality of reagents is critical. It is advisable to use high-quality, commercially sourced reagents and to regularly check their expiration dates to avoid potential issues related to contamination or nonspecific reactions.
2. Contamination control is another vital aspect of the process. Establishing a designated workspace that is free from contaminants is crucial. This includes maintaining separate areas for pre-amplification and post-amplification activities. Using aerosol-resistant pipette tips and dedicated pipettes for RT-LAMP reactions can help mitigate the risk of cross-contamination. Regular decontamination of surfaces and equipment with suitable disinfectants, such as bleach, is also recommended.
3. Proper sample handling is essential to avoid introducing contaminants. Wearing gloves and changing them frequently can minimize this risk.
4. Utilizing sterile tools for sample collection and processing further ensures the integrity of the samples being analyzed.
5. It is important to optimize the conditions for each assay, as different viruses may have varying temperature and time requirements for effective detection.
6. Validating the assay with known positive and negative controls is also a best practice to confirm the reliability of results.
7. Proper training in interpreting the results is necessary to avoid false positives and negatives, and utilizing appropriate software or methods for result analysis can facilitate accurate interpretation.
8. Storage of samples and reagents should be done according to recommended guidelines to maintain their integrity, and repeated freeze-thaw cycles of RNA samples should be avoided.
9. Personal protective equipment (PPE), including lab coats, gloves, and eye protection, is essential for safety and to prevent contamination.

References

1. Notomi T, Okayama H, Masubuchi H, Yonekawa T, Watanabe K, Amino N, Hase T (2000) Loop-mediated isothermal amplification of DNA. Nucleic Acids Res 28(12):e63–e63
2. Mori Y, Nagamine K, Tomita N, Notomi T (2001) Detection of loop-mediated isothermal amplification reaction by turbidity derived from magnesium pyrophosphate formation. Biochem Biophys Res Commun 289(1):150–154
3. Bhat AI, Rao GP (2020) Characterization of plant viruses, Springer protocols handbooks, vol 23. Humana, New York
4. Bhat AI, Siljo A, Deeshma KP (2013) Rapid detection of piper yellow mottle virus and

cucumber mosaic virus infecting black pepper (Piper nigrum) by loop-mediated isothermal amplification (LAMP). J Virol Methods 193(1):190–196

5. Nagamine K, Hase T, Notomi TJMCP (2002) Accelerated reaction by loop-mediated isothermal amplification using loop primers. Mol Cell Probes 16(3):223–229

6. Almasi MA, Aghapour-Ojaghkandi M, Aghaei S (2013) Visual detection of curly top virus by the colorimetric loop-mediated isothermal amplification. J Plant Pathol Microb 4:198

7. Nie K, Qi SX, Zhang Y, Luo L, Xie Y, Yang MJ et al (2012) Evaluation of a direct reverse transcription loop-mediated isothermal amplification method without RNA extraction for the detection of human enterovirus 71 subgenotype C4 in nasopharyngeal swab specimens. PLoS One 7(12):e52486

8. Bhat AI, Aman R, Mahfouz M (2022) Onsite detection of plant viruses using isothermal amplification assays. Plant Biotechnol J 20(10): 1859–1873

Chapter 29

Recombinase Polymerase Amplification for Viral Detection

Lubna Javaid, Sumiah Wani, Nulevino Iralu, Shahjahan Rashid, Sahar Saleem, and Aflaq Hamid

Abstract

Recombinase polymerase amplification (RPA) offers a rapid, sensitive, and isothermal approach for detecting viral infections in plants. Unlike traditional PCR methods, RPA operates at a constant low temperature of approximately 37 °C, eliminating the need for sophisticated thermal cyclers. This study outlines the application of RPA for the detection of plant viruses, highlighting its efficiency and simplicity. Plant tissue samples were collected and viral RNA/DNA was extracted using a plant nucleic acid extraction kit. The RPA reaction was prepared with specific primers for the target virus, along with the necessary enzymes and buffers. For RNA viruses, a reverse transcriptase enzyme was included. The reactions were incubated at 37–42 °C for 20–40 min, depending on the viral target. Detection of the amplified products was achieved through various methods, including agarose gel electrophoresis, lateral flow assays, and fluorescence-based detection systems. The results demonstrated that RPA is a versatile and effective tool for the rapid detection of plant viruses, providing a valuable alternative to conventional molecular diagnostic techniques.

Key words Recombinase polymerase amplification, Detection, Thermal cycler, Gel electrophoresis, Fluorescence

1 Introduction

Recombinase polymerase amplification (RPA) is an innovative isothermal DNA amplification and detection technology [1]. Unlike PCR, which requires a heat denaturation step at 95 °C, RPA uses a mixture of enzymes, including polymerase and DNA recombination proteins, to achieve DNA amplification. Specifically, RPA employs *Escherichia coli* RecA recombinase and single-strand DNA-binding protein (SSB) to denature DNA.

During the process, a phage-derived recombinase binds to single-stranded oligonucleotide primers, forming a stable helical filament. This filament efficiently scans double-stranded DNA (dsDNA) templates to identify homologous sequences within the target DNA. The recombinase proteins facilitate strand invasion and the formation of a D-loop structure at the target site, allowing

Aflaq Hamid et al. (eds.), *Detection of Plant Viruses: Advanced Techniques*, Springer Protocols Handbooks, https://doi.org/10.1007/978-1-0716-4390-7_29,

the primers to anneal. The complementary strand is then stabilized by SSBs, and the recombinase disassembles from the nucleoprotein filament, a process driven by ATP hydrolysis and the action of RecA protein.

DNA polymerase from *Staphylococcus aureus* (Sau DNA polymerase), which has strand displacement activity, initiates synthesis from the primer, creating new DNA strands. These newly synthesized strands serve as templates for subsequent rounds of amplification, resulting in the exponential amplification of the target DNA [2, 3] in less than 30 min [1, 4–6].

RPA reactions can be conducted at a single, constant low temperature of around 37 °C, eliminating the need for sophisticated thermal cyclers [1, 7–9]. These reactions can be completed in 5–20 min, depending on the initial copy number of the targeted templates and the size of the amplicon [10]. For RNA targets, a reverse transcriptase enzyme can be added to the reaction mixture, which is then performed at 40–42 °C, allowing for target detection in about 20 min [11].

RPA products can be detected using various methods:

- *Agarose gel electrophoresis*: RPA products can be visualized on an agarose gel [1], but the reaction mixture needs to be purified first to prevent protein interference that can cause smearing.
- *TwistAmp™ Exo probes* [9, 12]: These allow for quantitative detection using fluorescence and can be monitored in real time with a fluorescent detector, typically yielding results in 10–12 min.
- *Lateral flow dipstick assay*: Using a dipstick assay (e.g., from MileniaBiotec), results can be obtained simply and quickly.

The use of RPA for amplifying nucleic acids from bacteria and viruses has been well-documented, showcasing its versatility and efficiency for rapid and sensitive nucleic acid detection [13].

2 Materials

1. *RPA Kit*: Obtain a commercially available RPA kit (e.g., TwistDx or similar).
2. *Primers*: Specific primers for the viral target (*see* **Note 1**).
3. *Template DNA/RNA*: Extracted viral nucleic acid (*see* **Note 2**).
4. *Buffer*: Provided in the RPA kit.
5. *Magnesium Acetate*: Provided in the RPA kit.
6. *Nuclease-free water*: For dilutions and reactions.
7. *Detection reagents*: Such as lateral flow strips or fluorescence-based detection systems.

8. *Heat block or water bath*: Maintained at 37–42 °C.
9. *Pipettes and tips*: For accurate measurement and transfer of reagents.
10. *Microcentrifuge tubes*: For reaction mixtures.
11. *Optional*: RNAse inhibitor if working with RNA viruses.

3 Methods

3.1 Preparation of Reaction Mix

1. Thaw all reagents on ice and gently mix by pipetting up and down (*see* **Note 3**).
2. Prepare the RPA master mix according to the manufacturer's instructions, typically as follows:
1. X μL RPA rehydration buffer.
2. Y μL forward primer (10 μM).
3. Y μL reverse primer (10 μM).
4. Z μL template DNA/RNA (adjust based on template concentration).
5. Nuclease-free water to bring the total volume to the desired amount (usually 50 μL).

3.2 Initiation of the Reaction

1. Add the prepared master mix to a microcentrifuge tube containing the lyophilized RPA enzymes (from the kit) (*see* **Note 4**).
2. Add 2.5 μL of 280 mM magnesium acetate to initiate the reaction.
3. Mix by gently flicking the tube or pipetting up and down.

3.3 Incubation

1. Incubate the reaction mix at 37–42 °C for 20–40 min. The exact time may vary depending on the specific kit and target virus (*see* **Note 5**).
2. During incubation, the reaction can be periodically mixed by gentle inversion or pipetting to ensure uniform amplification.

3.4 Detection of Amplified Product

1. Detection methods vary depending on the kit and the type of primers/probes used:
2. Lateral flow assay: Dip a lateral flow strip into the reaction mixture and read the result according to the manufacturer's instructions.
3. Fluorescence detection: If using fluorescent probes, measure the fluorescence signal using an appropriate device.

4. Gel electrophoresis: If applicable, run a small aliquot of the amplified product on an agarose gel and visualize under UV light.

3.5 Interpretation of Results

1. Compare the results with positive and negative controls to determine the presence of the viral target.
2. A positive result indicates the presence of the viral DNA/RNA in the sample.

4 Notes

1. Primer Design: Ensure primers are specific to the viral target and do not form secondary structures.
2. Template Quality: Use high-quality, purified nucleic acids to avoid inhibition of the reaction.
3. Contamination Control: Use separate areas and equipment for reagent preparation and template handling to prevent contamination.
4. By following this protocol, you should be able to detect viral nucleic acids efficiently using the RPA method. Always refer to the specific instructions provided by the RPA kit manufacturer for best results.
5. Reaction Conditions: Optimize incubation time and temperature if necessary for better results.

References

1. Piepenburg O, Williams CH, Stemple DL, Armes NA (2006) DNA detection using recombination proteins. PLoS Biol 4(7):e204
2. West SC (2003) Molecular views of recombination proteins and their control. Nat Rev Mol Cell Biol 4(6):435–445
3. Krejci L, Altmannova V, Spirek M, Zhao X (2012) Homologous recombination and its regulation. Nucleic Acids Res 40(13): 5795–5818
4. Shibata T, Cunningham RP, DasGupta C, Radding CM (1979) Homologous pairing in genetic recombination: complexes of recA protein and DNA. Proc Natl Acad Sci 76(10): 5100–5104
5. Yonesaki T, Ryo Y, Minagawa T, Takahashi H (1985) Purification and some of the functions of the products of bacteriophage T4 recombination genes, uvs X and uvs Y. Eur J Biochem 148(1):127–134
6. Armes NA, Stemple DL (2007) U.S. Patent No. 7,270,981. Washington, DC: U.S. Patent and Trademark Office
7. Hoff M (2006) DNA amplification and detection made simple (relatively). PLoS Biol 4(7): e222
8. Euler M, Wang Y, Nentwich O, Piepenburg O, Hufert FT, Weidmann M (2012) Recombinase polymerase amplification assay for rapid detection of Rift Valley fever virus. J Clin Virol 54(4):308–312
9. Euler M, Wang Y, Otto P, Tomaso H, Escudero R, Anda P et al (2012) Recombinase polymerase amplification assay for rapid detection of Francisella tularensis. J Clin Microbiol 50(7):2234–2238
10. Hill-Cawthorne GA, Hudson LO, El Ghany MFA, Piepenburg O, Nair M, Dodgson A et al (2014) Recombinations in staphylococcal cassette chromosome mec elements compromise the molecular detection of methicillin

resistance in Staphylococcus aureus. PLoS One 9(6):e101419

11. Abd El Wahed A, Weidmann M, Hufert FT (2015) Diagnostics-in-a-Suitcase: Development of a portable and rapid assay for the detection of the emerging avian influenza A (H7N9) virus. J Clin Virol 69:16–21
12. Boyle DS, Lehman DA, Lillis L, Peterson D, Singhal M, Armes N et al (2013) Rapid detection of HIV-1 proviral DNA for early infant diagnosis using recombinase polymerase amplification. MBio 4(2):10–1128
13. Zhang S, Ravelonandro M, Russell P, McOwen N, Briard P, Bohannon S, Vrient A (2014) Rapid diagnostic detection of plum pox virus in Prunus plants by isothermal AmplifyRP® using reverse transcription-recombinase polymerase amplification. J Virol Methods 207:114–120

Chapter 30

Detection of Plant Viruses by CRISPR-Cas13a System

Mariya Rather, Sumiah Wani, Shahjahan Rashid, Nulevino Iralu, Anirban Roy, Hanu R. Pappu, and Aflaq Hamid

Abstract

Viral infections in plants are a serious threat to food security. Current viral diagnostic methods typically PCR are effective but have numerous limitations, like the need for sophisticated equipments in a laboratory setting as well as the requirement for skilled technical staff, besides being time consuming. It has, therefore, become imperative to develop novel diagnostic tools that are fast, sensitive, have high specificity, are cost-effective, and need minimal sample preparation and equipment, particularly ones that can be used for versatile point-of-care diagnostic applications. Such field-deployable methods for nucleic acid detection are crucial to prevent extensive crop losses due to viral diseases. A CRISPR-Cas-based diagnostic platform termed "specific high-sensitivity enzymatic reporter unlocking" (SHERLOCK) has been developed that allows portable, ultrasensitive, multiplexed, and rapid detection of nucleic acids. The CRISPR-Cas13a diagnostic tool combines isothermal pre-amplification of nucleic acids with CRISPR-Cas enzymology for rapid detection of desired DNA or RNA sequences. Compared to the traditional diagnostic methods, CRISPR-Cas13a-based detection systems are able to simultaneously achieve high sensitivity, high specificity, and a low-cost detection of target nucleic acids, which makes it extremely useful in a low-resource setting.

Key words Nucleic acid detection, CRISPR diagnostics, CRISPR-Cas13a, SHERLOCK, Plant virus

1 Introduction

Plant virus diagnostics have evolved significantly, resulting in more accurate and faster detection. The gold standard for virus detection has traditionally involved comprehensive molecular techniques based on the amplification of the viral genome by polymerase chain reaction (PCR) and its derivative approaches [e.g., reverse transcription followed by quantitative PCR (RT-qPCR)], or to detect viral proteins by the enzyme-linked immunosorbent assay (ELISA) [1]. These methods are effective but require sophisticated instrumentation in a laboratory setting, precluding their deployment in the field. Demand for instrument-free nucleic acid detection technologies has driven the development of multiple methods

Aflaq Hamid et al. (eds.), *Detection of Plant Viruses: Advanced Techniques*, Springer Protocols Handbooks, https://doi.org/10.1007/978-1-0716-4390-7_30,

for isothermal amplification. These methods, which include recombinase polymerase amplification (RPA) or loop-mediated isothermal amplification (LAMP), are fast, sensitive, and require minimal sample preparation and equipment; therefore, are useful in field-based scenarios. However, these techniques require optimization and lack high specificity [2]. Recently, enzymes from CRISPR-Cas systems have been adapted for the specific, rapid, sensitive, and portable detection of nucleic acids [3]. These methods rely on Cas13 [4] or Cas 12 enzymes [5], both of which exhibit nonspecific endonuclease activity after binding to a specific target via programmable crRNAs. A rapid and easy nucleic-acid detection system based on CRISPR-Cas13a, termed "specific high-sensitivity enzymatic reporter unlocking" (SHERLOCK), that allows multiplexed, portable, and ultrasensitive detection of RNA has become a highly useful diagnostic tool that combines nucleic acid pre-amplification with CRISPR-Cas enzymology for specific recognition of desired RNA sequences [6].

Cas13 is an RNA-guided RNase that produces multiple cleavage sites in a single-stranded RNA target with specific base preferences. Additionally, Cas13 demonstrates target-dependent promiscuous RNAase activity, which causes bystander RNA molecules to cleave, a phenomenon known as "collateral activity." On recognition of its RNA target, activated Cas13a engages in "collateral" cleavage of nearby nontargeted RNAs [4]. SHERLOCK combines isothermal amplification that includes recombinase polymerase amplification (RPA) and T7 transcription with the collateral cleavage effect of Cas13a for virus detection [7]. The detection of the amplified targets can be achieved with the aid of a variety of reporter molecules and methods, including (i) fluorometer detection with quenched fluorescent reporters [7, 8] and (ii) lateral flow detection assays [9]. In case of the fluorescent method, detection of target RNAs is achieved by introducing RNA fluorescent reporter molecules that are connected to both fluorescence and quenching groups. The activated Cas13a not only cleaves the target RNA but due to its collateral activity cleaves the nearby RNA fluorescent reporter molecules away from the quencher. A detectable fluorescent signal is thus released which can be read by a fluorometer [8]. For simpler instrument-free and portable detection, lateral flow readouts can be used [9].

2 Materials

1. Plant extraction buffer (500 mM NaOH, 10 Mm EDTA).
2. UltraPure distilled water.
3. Murine RNase inhibitor.
4. RNaseAway.

5. TwistAmp Basic kit (with TwistAmp rehydration buffer, TwistAmp magnesium acetate).
6. HEPES buffer.
7. Magnesium chloride ($MgCl_2$, 0.2 μm filtered).
8. LwaCas13a in SB (Super Broth).
9. RNaseAlert v2 Reporter (2 μM).
10. T7 RNA polymerase.
11. Water bath.
12. Constant temperature incubator.
13. Culture flasks.
14. Fluorescence plate reader.

3 Method

Nucleic acid detection by SHERLOCK includes the following steps: (i) sample extraction, (ii) isothermal target nucleic acid pre-amplification, and (iii) CRISPR-Cas13 nucleic acid detection (Fig. 1).

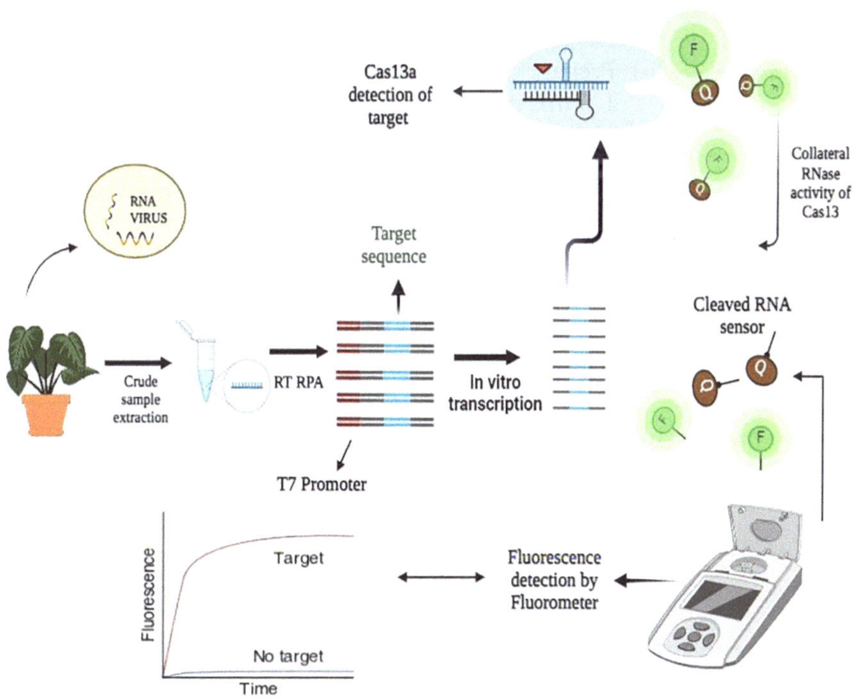

Fig. 1 Schematic representation of CRISPR-Cas13a detection of RNA virus by Fluorescence-based assay (Chapter 30)

3.1 Sample Extraction: (Time: 5 Min to 1 h)

SHERLOCK detection is compatible with both pure and crude sample extractions, as long as contaminating nucleases are inactivated. Nuclease carryover to SHERLOCK reactions may degrade the reporter and lead to false-positive background signals. The addition of RNase inhibitors is therefore important.

3.2 Crude Extraction from Plant Material

1. Add 200 μL of plant extraction buffer (500 mM NaOH, 10 Mm EDTA) to 20 mg of crushed plant sample (seeds, leaf, or root tissue).
2. Vortex the sample for 30 s and incubate at room temperature for 1 min. The crude alkaline extracted product can be stored at −20 ° C for up to 2 weeks.
3. Transfer a 5 μL aliquot to a new tube and dilute 1: 20 with 95 μL of UltraPure water. Only 1 μL of sample is needed for the isothermal pre-amplification step. The remaining dilution can be stored at −20 °C for up to 2 weeks.

3.3 RPA-Based Pre-amplification of Target DNA/RNA (Time 30 Min to 1 h)

SHERLOCK involves a pre-amplification step, typically RPA (recombinase polymerase amplification), that amplifies a DNA or RNA target. Thereafter, amplified targets are converted to RNA via T7 transcription followed Cas13 detection.

STEPS:

(i) Wipe the work area and pipettes with RNaseAway for eliminating any RNase contamination (*see* **Note 1**).

(ii) Prepare the following reaction master-mix on ice in a 1.5 mL tube:

Component	Volume (μL)
Forward primer	2.4
Reverse primer	2.4
TwistAmp rehydration Buffer (from TwistAmp basic kit)	29.5
UltraPure water	8.65

(iii) Add 40 μL of the master-mix to a reaction pellet. Pipette up and down until the pellet is resuspended.

(iv) Add 2.5 μL of 280 mM TwistAmp magnesium acetate (from the TwistAmp Basic kit) to the reaction. Vortex briefly and do a short spin to make the magnesium acetate drop fall down into the tube (*see* **Note 2**).

(v) Prepare aliquots of 10 μL of the RPA reaction in PCR tubes, and add 1 μL of sample to each tube. Mix by carefully by pipetting up and down.

(vi) Incubate at 37 °C for 10–30 min (*see* **Note 3**).

3.4 Cas13 Nucleic Acid Detection (Time: 1–4 h)

Fluorescence-based detection assay:

1. Preheat the fluorescence plate reader to 37 °C.
2. Prepare the Cas13-SHERLOCK master mix by adding the following components to an Eppendorf tube.

Component	Vol per reaction (μL)	Vol for four replicates (μL)
UltraPure water	11.27	56.35
HEPES buffer, pH 6.8, 1 M	0.4	2
MgCl2, 1 M	0.18	0.9
rNTP solution mix, 25 mM each	0.8	4
LwaCas13a in SB, 63.3 μg/mL	2	10
Murine RNase inhibitor, 40 U/μL	1	5
T7 RNA polymerase, 5 U/μL	0.5	2.5
crRNA (10 ng/μL)	1	5
RNaseAlert v2 reporter (2 μM)	1.25	6.25
Total	19	95

3. For four replicates, transfer 87.4 μL of the master-mix to a PCR strip tube or 96-well PCR plate.
4. Spin down the RPA pre-amplification reaction sample.
5. Add 4.6 μL of the RPA reaction mix to the 87.4 μL of the master-mix on ice. Briefly vortex and spin down (using a mini centrifuge or plate spinner for 5 s) in a 96-well plate to collect the entire reaction mix in the well.
6. Open each Eppendorf tube and transfer 20 μL per technical replicate and condition to a 384-well, round black-well, clear-bottom plate on ice.
7. Briefly spin down the plate (@500 g, 22 °C, 15 s) to remove any bubbles and place it into a preheated fluorescence plate reader (*see* **Note 4**).
8. The fluorescence signal data can be read by a fluorometer.

4 Notes

1. It is important for eliminating any RNase contamination.
2. Addition of magnesium acetate will make the reaction start. It is therefore important to quickly proceed to sample addition to avoid primer-dimer formation.
3. After incubation, the test samples could be either immediately run for Cas13 detection or can also be stored at 4 °C for 2–3 days. For longer storage period, reaction should be stored at −20 °C.
4. After plate preparation, samples should be immediately placed onto the fluorescence plate reader, because SHERLOCK Cas13 reaction will start as the samples are removed from ice.

References

1. Rubio L, Galipienso L, Ferriol I (2020) Detection of plant viruses and disease management: relevance of genetic diversity and evolution. Front Plant Science 11:1092
2. Marques CM, Vicente SJ, Ruiz R, Martinez MR, Costa MR, Gomez G, Carbonell A, Daros AJ, Rodrigo G (2022) Diagnostics of infections produced by the plant viruses TMV, TEV and PVX with CRISPR-Cas12 and CRISPR-13. ACS Synth Biol 11:2384–2393
3. Bhattacharyya RP, Thakku SG, Hung DT (2018) Harnessing CRISPR effectors for infectious disease diagnostics. ACS Infect Dis 4: 1278–1282
4. Abudayyeh OO, Gootenberg SJ, Konermann S, Joung J, Slaymaker MI, Cox BTD, Shmakov S, Makarova SK, Semenova E, Minakhin L, Severinov K, Regev A, Lander SE, Koonin VE, Zhang F (2016) C2c2 is a single-component programmable RNA-guided RNA-targeting CRISPR effector. Science 353:6299
5. Zetsche B, Gootenberg SJ, Abudayyeh OO, Slaymaker MI, Makarova SK, Essletzbichler P, Volz ES, Joung J, Van der Oost J, Regev A, Koonin VE, Zhang F (2015) Cpf1 is a single RNA-guided endonuclease of a class 2 CRISPR-Cas system. Cell 163:759–771
6. Kellner MJ, Koob JG, Gootenberg SJ, Abudayyeh OO, Zhang F (2020) SHERLOCK: nucleic acid detection with CRISPR nucleases. Nat Protoc 15:131
7. Gootenberg SJ, Abudayyeh OO, Lee WK, Essletzbichler P, Aaron JD, Joung J et al (2017) Nucleic acid detection with CRISPR-Cas13a/C2c2. Science 356(6336):438–442
8. East-Seletsky A, O'Connell MR, Knight SC, Burstein D, Cate JH, Tjian R, Doudna AJ (2016) Two distinct RNase activities of CRISPR-C2c2 enable guide-RNA processing and RNA detection. Nature 538:270–273
9. Gootenberg SJ, Abudayyeh OO, Kellner JM, Joung J, Collins JJ, Zhang F (2018) Multiplexed and portable nucleic acid detection platform with Cas13, Cas12a, and Csm6. Science 360: 439–444

Chapter 31

Detection of Plant Viruses Through RNA-Seq

Dasari Meghanath, Sumiah Wani, Kottam Sushma, Rao Thulasi, Nulevino Iralu, and Aflaq Hamid

Abstract

RNA sequencing (RNA-seq) has revolutionized plant virus detection. This powerful technique offers an unbiased approach to identify a wide range of viruses, including those previously unknown. Unlike traditional methods targeting specific pathogens, RNA-seq analyzes the entire pool of RNA molecules within a plant sample. This comprehensive approach allows for the simultaneous detection of multiple viruses, even in mixed infections and also collects the host transcriptome data. In RNA-Seq experiment, removal of ribosomal RNA (rRNA) to enrich for viral RNA sequences is critical as these are overshadowed with mRNA. However, RNA-seq can detect viruses with varying genome types, including those with and without polyadenylated tails. Beyond detection, RNA-seq has the ability to not only identify viruses but also to explore host plant responses to infection through analysis of the host transcriptome data obtained along with viral sequences. This method briefly explains the careful depletion of ribosomal RNA (Sd rRNA) followed by RNA sequencing library preparation to detect a wide range of viruses.

Key words RNA-Seq, Viruses, Transcriptome, Selective depression of rRNA, RNA sequencing library

1 Introduction

Traditional methods for virus detection often target specific known viruses in a combination of approaches, identifying specific viral proteins with antibodies in an immunological test, direct visualization of virus particles under an electron microscope, and utilizing nucleic acid probes to detect the viral genetic material [1]. These detection methods, including those using specific nucleic acid probes, can struggle when dealing with rapidly evolving viruses. Mutations in the targeted viral sequence and increase in the number of strains can render these probes/primers ineffective, potentially leading to missed diagnoses. High-throughput RNA sequencing (RNA-seq) provides a significant advantage by offering a more comprehensive analysis. By sequencing all the RNA in a sample, RNA-seq can identify the complete set of viruses present, including new viruses that cannot be identified through traditional methods.

Aflaq Hamid et al. (eds.), *Detection of Plant Viruses: Advanced Techniques*, Springer Protocols Handbooks,
https://doi.org/10.1007/978-1-0716-4390-7_31,

However, RNA-seq offers an unbiased approach. It analyzes all RNA molecules present in a sample and enables to identify *both known and unknown* viruses infecting host plants. This is crucial because new viruses can emerge and cause significant damage before they are even identified. Thus, RNA sequencing (RNA-seq) turned up as a powerful tool for detection of plant viruses, offering a two-pronged approach. It allows to uncover and identify novel plant viruses by detecting all types of RNA viruses present in the infected sample and also reveals the expression of host plant genes in response to viral infection, allowing for parallel collection of host plant transcriptome data. rRNA removal is critical for accurate RNA-seq data when detecting viruses using RNA-seq. Since rRNA doesn't provide understanding into gene expression, its presence can obscure the signals from the mRNA of interest. Removing rRNA through an appropriate method from total RNA ensures a clear picture of the transcriptome and allows to focus on the most relevant RNA molecules [2]. While RNA-Seq with oligo-dT bead purification excels at detecting viruses with polyadenylated (poly(A)) tails, it has limited effectiveness for those viruses lacking them. However, the rRNA selective depression method within RNA-Seq offers a powerful solution. This approach bypasses the poly(A) requirement, enabling comprehensive detection of a wider range of viral genomes, including those without poly(A) tails [2]. This approach has led to the identification of several *putative viruses* infecting cultivated crops, jumping the species barrier and establishing themselves in new hosts. In this chapter, an RNA sequencing protocol utilizing selective ribosomal RNA (sRNA) depletion and optimized RNA-seq library preparation for comprehensive transcriptome analysis was described, which was mostly based on Nagano et al. [3], Wang et al. [4], Zong et al. [5].

2 Materials

2.1 Collection of Samples

1. Polythene lock covers or RNA later solution or Liquid nitrogen.

2.2 RNA Isolation

1. RNeasy Plant Mini Kit (QIAGEN).
2. TRIZOL reagent.
3. Chloroform.
4. Isopropanol.

2.3 RNA Quantity and Quality Assessment

1. Nanodrop spectrophotometer (Thermofisher scientific).

2.4 Selective Depression of rRNA (SDRNA)

1. 0.4 μM each in final concentration of 100 μM SDRNA oligo pool.
2. ERCC RNA Spike-In Mix (Synthetic RNA molecules).
3. Hybridization buffer (10× concentration) containing Tris–HCl (1 M, pH 7.4) and NaCl (2 M).
4. Ribonuclease H (10×) digestion buffer: Tris–HCl (500 mM, pH 7.4), NaCl (1 M), and $MgCl_2$ (200 mM).
5. Thermostable Ribonuclease H (Specifically degrade the RNA in a DNA-RNA hybrid).
6. AMpure XP beads (Beckman Coulter, for purification of PCR products).
7. DNase I free from RNase.
8. Dithiothreitol (100 mM, DTT, to stabilize enzymes and other proteins).
9. AMpureXP buffer: Polyethylene glycol (20% PEG) and NaCl (2.5 M).

2.5 RNA-Seq Library Preparation

1. Random hexamer primer (Anneals at multiple points).
2. Heat stable RNasin Plus (RNase inhibitors).
3. 25 mM each dNTP.
4. MMLV Reverse Transcriptase (MMLV RTase).
5. Absolute ethanol.
6. 2 mM dATP, dCTP, dGTP, and 4 mM dUTP mix.
7. Ribonuclease H (*Endonuclease*).
8. DNA polymerase I (Enzymatics).
9. 2.5 mM each dNTP.
10. End-Repair Mix (Produces blunt end 5^1 DNA).
11. 10 mM dATP mix (Promega).
12. Klenow fragment $3' \rightarrow 5'$ exonuclease (Fills 5^1 overhangs).
13. DNA Ligase.
14. Y-shape adapter (Multiple annealing sequences).
15. 12P beads (*see* **Note 1**).
16. Uracil DNA glycosylase (UNG or UDG).
17. Index primer (*see* **Note 2**).
18. Universal primer.
19. 5× Phusion HF Reaction buffer (50 mM MgCl2 and DMSO).
20. Phusion DNA polymerase.
21. E-Gel Power snap electrophoresis system.
22. E-Gel EX with SYBR Gold SizeSelect II Agarose Gels, 2%.

3 Methods

3.1 Collection of Samples

Field samples should be collected properly in replicates and should be stored in liquid nitrogen immediately after collection or in an RNA later solution or store at −80 °C in a deep freezer until RNA isolation (*see* **Note 3** and **4**).

3.2 RNA Isolation

Total RNA can be isolated using TRIZOL reagent (Invitrogen) as described in previous chapter (*see* **Note 5**) or RNEasy Plant Mini kit (Qiagen) as per the instructional manual.

3.3 RNA Quality

The RNA quality and quantity can be assessed (*see* **Note 6**) using spectrophotometer (Thermofisher Scientific).

3.4 SDRNA (Depression of Ribosomal RNA)

1. Mix 500 nanograms (ng) of total RNA sample, a recommended volume of ERCC RNA Spike-In control mix (manufacturer's instructions for specific volume), 1 microliter (μL) of the SDRNA oligo pool, RNase-free water to a final volume of 10 microliters (μL).
2. Place the mixture in a thermal cycler and run the following: Initial denaturation at 95 °C for 2 min, slow annealing at 45 °C, hybridization hold at 45 °C for 5 min.
3. Prepare and add a pre-warmed premix to the above combining 1 μL thermostable RNaseH, 2 μL RNaseH digestion buffer (10×), and 7 μL RNase-free water, mix gently by pipetting.
4. *Incubate the reaction mixture at 45 °C for 30 min.* This incubation allows the Hybridase Thermostable RNaseH enzyme to selectively digest the RNA portion of the DNA/RNA hybrid double strand.
5. Add 24 microliters (μL) of AMpure XP beads to the reaction mixture in the PCR tube. *Mix the solution thoroughly by pipetting up and down several times.*
6. Incubate the mixture at room temperature for the recommended time (refer to AMpure XP beads protocol you use).
7. *Place the tube on a magnetic stand and incubate for another minute to allow the beads to pellet.* The solution should become clear as the beads migrate to the side of the tube.
8. *Carefully remove and discard the supernatant without disturbing the beads.*
9. Resuspend beads with 10 microliters (μL) of RNase-free water for further processing.
10. Prepare DNaseI premix (*see* **Note 7**) solution to remove DNA contamination from the RNA sample by combining 1 μL of DNaseI enzyme (5 U), 1 μL of 10× DNaseI buffer, 1 μL of 100 mM DTT (Dithiothreitol, a reducing agent), and 7 μL of RNase-free water.

11. The DNaseI premix is added to the resuspended beads and incubated at 37 °C for 30 min. This allows DNaseI to degrade any DNA present in the sample.
12. Add 24 μL of AMpure XP buffer to the sample and perform purification using a magnetic stand. This step isolates the purified RNA using magnetic beads.
13. The AMpure XP magnetic beads containing the purified RNA should be used immediately in the next step to avoid over-drying. Over-drying can reduce RNA yield or quality.

3.5 RNA Sequencing Library Preparation

1. Prepare a master mix for RNA fragmentation with 2 μL MMLV buffer (*see* **Note 8**), DTT (1 μL), and RNase free water (7 μL).
2. Carefully mix the beads containing the purified RNA with the fragmentation master mix and heat at 94 °C for 5 min to cut the RNA molecules into smaller pieces. Cool the mixture immediately down to 4 °C to inactivate the fragmentation enzymes and to prevent further RNA degradation.
3. Place the tube containing the cooled mixture on a magnetic stand till the supernatant containing the fragmented RNA becomes clear and transfer the supernatant to a new tube.
4. Add random hexamer (2 μL) and RNasin Plus (1 μL) to 10 μL of fragmented RNA solution and incubate at 50 °C for 5 min to anneal to the fragmented RNA solution and quench it on ice immediately.
5. Prepare the RT master mix by combining 1 μL DTT, 0.1 μL dNTP, 0.5 μL MMLV RTase (reverse transcriptase enzyme), and 5.9 μL of RNase-free water.
6. The fragmented RNA solution with random hexamers and RT master mix are combined and incubated at 25 °C for 10 min and then for 50 min at 42 °C.
7. AMpure XP beads (magnetic beads) are used to purify the newly synthesized cDNA along with the remaining fragmented RNA (referred to as cDNA/RNA hybrid solution).
8. Add 36 μL of AMpure XP beads and 18 μL of 100% ethanol to the reaction mixture. Refer to manufacturer's instructions for detailed purification steps using the magnetic stand.
9. After purification, the beads are washed, and the cDNA/RNA hybrid solution is eluted (removed) by resuspending the washed beads with 10 μL of nuclease-free water.
10. Finally, the supernatant (liquid portion) containing the purified cDNA/RNA hybrid solution is transferred to a new tube using the magnetic stand. This purified product is likely used in further downstream applications.

11. Prepare the master mix for the synthesis of the second strand using 10× Blue Buffer (2 μL) (*see* **Note 9**), dUTP/NTP mix (1 μL), 100 mM DTT (0.5 μL), RNase H (0.5 μL), DNA polymerase I (1 μL), and nuclease-free water (5 μL).
12. Mix 10 μL of the second strand synthesis master mix with 10 μL of the purified DNA/RNA hybrid solution, excluding beads.
13. After incubating overnight at 16 °C, proceed by adding 25 μL of AMpure XP beads and purifying the mixture using a magnetic stand.
14. Suspend the washed beads again in 10 microliters of water that is free from nucleases.
15. Mix the following to create the end-repair master mix (*see* **Note 10**): Concentrated End Repair Buffer (2 μL), dNTP mix (1 μL with each component at 2.5 mM), End-Repair Mix LC (1 μL), and NFW (6 μL).
16. Add 10 μL of the prepared end-repair master mix to the 10 μL of DNA solution containing beads, which has been previously purified.
17. Allow the mixture to incubate for 30 min at 20 °C.
18. Purify the mixture using the magnetic stand, after adding 25 μL of AMPure XP buffer.
19. Resuspend the washed beads by adding 10 μL of NFW.
20. Combine 2 μL of Blue Buffer, 1 μL of dATP, 0.5 μL of Klenow enzyme, and 6.5 μL of nuclease-free water to create the dA-tailing master mix. Then, add this mix to the purified DNA solution containing beads, in a 1:1 ratio (10 μL each), to perform the dA-tailing reaction.
21. Place the mixture in an incubator at a temperature of 37 °C for a duration of 30 min.
22. Mix in 25 μL of AMPure XP buffer, followed by purification using the magnetic stand. This process will isolate the targeted components via magnetic separation.
23. Suspend the washed beads again in 10 μL of nuclease-free water, mixing vigorously to ensure a consistent mixture.
24. Separate the A-tailed DNA solution from the magnetic beads using the magnetic stand, and transfer the supernatant to a fresh tube.
25. Mix 1 μL of the 0.1 μM Y-shape adapter with 10 μL of the A-tailed DNA solution, creating a combined solution.
26. Mix 12 μL of 2× Rapid Ligation Buffer (*see* **Note 11**) with 1 μL of T4 DNA Ligase to form a ligation master mix.

27. Pipette the ligation master mix, universal adapter, and A-tailed DNA solution into a single tube, mix thoroughly to ensure a homogeneous solution.
28. Place the mixture in an incubator at 20 °C for 20 min.
29. Mix 15 μL of 12P beads with 12 μL of the ligation product, place it on magnetic stand to capture the beads and purify the solution.
30. Suspend the washed beads again in 10 μL of nuclease-free water.
31. Add 10 μL of AMPure XP buffer to the 10 microliters of purified DNA.
32. Perform a second round of purification using the magnetic stand. Following purification, resuspend the beads in 15 μL of 10 mM Tris–HCl (pH 8.0) and mix well. Then, use the magnetic stand to capture the beads and transfer the supernatant (purified ligation product) to a fresh tube.
33. Add 1 μL of uracil DNA glycosylase to the 15 μL of purified ligation product and incubate at 37 °C for 30 min, enabling the enzyme to perform the necessary modifications on the DNA.
34. Mix 5.4 μL of UDG-digested DNA, 1 μL of index primer (10 μM), 1 μL of universal primer (10 μM), 2 μL of Phusion HF buffer (5x concentration), 0.3 μL of dNTPs (10 mM), and 0.3 μL of Phusion DNA polymerase to form a uniform solution.
35. Perform PCR amplification of DNA fragments with adapters and index sequences using a thermal cycler at 94 °C for 2 min, cycling (10–20 cycles), 98 °C for 10 s, 65 °C for 30 s, 72 °C for 30 s, and a final 72 °C for 5 min.
36. Mix 10 μL of AMpure XP beads with the PCR product. For multiple samples (N), pool the PCR products and add 10 × N μL of AMpure XP beads to the combined mixture, then proceed with the purification process.
37. Purify the mixture using the magnetic stand. Following purification, resuspend the beads in 25 μL of nuclease-free water. Then, use the magnetic stand to capture the beads and transfer the supernatant to a new tube, thereby isolating the purified material.
38. Use E-Gel SizeSelect Agarose Gels (2%) to perform size selection and remove fragments outside the desired range (200–600 bp).
39. Collect five fractions at 30-s intervals, beginning at 250 bp, and combine them into a single library. Store the library at −20 °C until sequencing. Prior to sequencing, evaluate the library's quality by running 1 μL of the combined library on the

Bioanalyzer. Next, the prepared library can undergo high-throughput sequencing using either the HiSeq or MiSeq platforms, both of which utilize Illumina's sequencing by synthesis (SBS) technology.

3.6 Mapping and Quantification

After sequencing, the summary and alignment statistics can be obtained through RSeQC. The poor-quality reads and adapter contamination can be removed through FASTQC and Trimmomatic. The QC passed reads can be mapped on to the reference genome or assembled de novo using bowtie or BWA.

4 Notes

1. 12P beads are prepared by removing buffer of AMpure XP and by adding equal volume of 12P buffer (12% PEG-8000, 2.5 M NaCl) to the removed AMpure XP buffer.
2. Index Primers are an index sequences of 80-mer to prepare amplicon libraries for sequencing on Illumina platforms by adding a unique, identifying "barcode" sequence to each amplicon. Use 96 sample indexes available for P5 and P7 Illumina index primer sequences.
3. After collecting plant samples from the field, try to store them immediately at −80° in a deep freezer in your lab or collect them in an RNA latter solution.
4. In case of unavailability of liquid nitrogen, label the samples properly and pack them in an air sealed polyethylene covers.
5. RNEasy plant mini kit yields quality RNA; however, in case of large number of plant samples for RNA isolation, TRIZOL method is suggested and the protocol was described in the previous chapter.
6. The Nanodrop spectrophotometer will typically display the absorbance values at each wavelength and calculate concentration (ng/μL) value derived from the absorbance at 260 nm using the Beer-Lambert Law. It tells you the amount of RNA in your sample per microliter. 260/280 ratio reflects the purity of your RNA. A ratio of around 2.0 is considered ideal for pure RNA. Lower values suggest protein contamination. 260/230 ratio above 2.0 indicates minimal organic or buffer contamination. Try to repeat the isolation, if your RNA quality deviates those values.
7. DNase I buffer contains 400 mM Tris–Hcl (pH 7.5), 50 mM DTT, and 80 mM $Mgcl_2$.
8. MMLV RT buffer contains 500 mM Tris–Hcl (pH 7.5), 100 mM DTT, 750 mM KCl, and 30 mM $Mgcl_2$.

9. 10× blue buffer contains 100 mM Tris–Hcl (pH 7.5), 10 mM DTT, 500 mM NaCl, and 100 mM $Mgcl_2$.
10. 10× end repair buffer contains 1 mM Tris–Hcl (pH 7.5), 50 mM DTT, 500 mM NaCl, 0.25% Triton-X 100, and 100 mM $Mgcl_2$.
11. The 2× rapid ligation buffer contains 130 mM Tris–HCl (pH 7.6), 2 mM DTT, 2 mM ATP, 20 mM $MgCl_2$, and 15% PEG 6000.

References

1. Hull R (2002) Matthews' plant virology, 4th edn. Elsevier, Amsterdam
2. Uyeda I, Masuta C (2015) Plant virology protocols: new approaches to detect viruses and host responses. Humana Press
3. Nagano AJ, Sato Y, Mihara M, Antonio BA, Motoyama R, Itoh H et al (2012) Deciphering and prediction of transcriptome dynamics under fluctuating field conditions. Cell 151:1358–1369
4. Wang L, Si Y, Dedow LK, Shao Y, Liu P, Brutnell TP (2011) A low-cost library construction protocol and data analysis pipeline for Illumina-based strand-specific multiplex RNAseq. PLoS One 6(10):e26426
5. Zhong S, Joung JG, Zheng Y, Chen YR, Liu B, Shao Y et al (2011) High-throughput illumina strand-specific RNA sequencing library preparation. In: Cold spring Harb protocols, vol 2011, pp 940–949

Chapter 32

Preparation of Sequencing Library for Oxford Nanopore Minion Sequencing

Nulevino Iralu, Dasari Meghanath, Sumiah Wani, Sahar Saleem, and Aflaq Hamid

Abstract

The ligation sequencing kit is a portable option for users looking to optimize their sequencing experiments for better output and control over read length. This kit can be used for upstream procedures such as size selection, whole genome amplification, or enrichment of long reads. This chapter highlights the operation protocol for preparing sequencing library for oxford minion sequencing.

Key words Sequencing, CpG sites, Genome, Ligation

1 Introduction

The ligation sequencing kit offers a versatile and portable solution for researchers aiming to enhance their sequencing experiments. By providing greater control over read length and output, this kit is particularly useful for various upstream processes. Whether size selection, whole genome amplification, or enriching for long reads is being performed, its adaptability makes it an excellent choice for optimizing sequencing workflows. This flexibility allows users to tailor their approach based on specific experimental needs, ultimately leading to more reliable and insightful results. Nanopore sequencing enables direct detection of methylated cytosines (CpG sites) without the need for bisulfite conversion [1]. Gene transcription and transposon activity is regulated by cytosine methylation in plant genomes. In mammalian cells, CpG sites occurs in density clusters called the CpG islands (CGI); however, plants do not exhibit clear CpG islands. Mutation of DNA methyltransferases for adaptation to environmental stresses or during development necessitates the genome-wide methylomes studies. There is need to monitor multiple samples for determination of transgenerational

Aflaq Hamid et al. (eds.), *Detection of Plant Viruses: Advanced Techniques*, Springer Protocols Handbooks,
https://doi.org/10.1007/978-1-0716-4390-7_32,

methylation inheritance or differential cytosine methylation [2]. Adaptive sampling (AS) offers a quick, flexible, and precise method for concentrating on specific regions of interest by minimizing off-target sequences during sequencing, without requiring prior sample preparation.

2 Materials

1. 2 μg extracted genomic DNA.
2. Ligation Sequencing Kit V14.
3. Flow Cell Wash Kit.
4. NEBNext FFPE Repair Mix (NEB, M6630).
5. NEBNext Ultra II End repair/dA-tailing Module (NEB, cat E7546).
6. NEBNext Quick Ligation Module (NEB, catE6056).
7. Freshly prepared 80% ethanol in nuclease-free water.
8. Nuclease-free water (e.g., ThermoFisher, AM9937).
9. 1.5 mL Eppendorf DNA LoBind tubes.
10. 0.2 mL thin-walled PCR tubes.
11. Qubit™ Assay Tubes (Invitrogen, Q32856).
12. Qubit dsDNA HS Assay Kit (Invitrogen, Q32851).
13. (Optional) Bovine Serum Albumin (BSA)(50 mg/mL)(e.g., Invitrogen™ UltraPure™ BSA 50 mg/ml,AM2616).
14. Hula mixer (gentle rotator mixer).
15. Magnetic separator, suitable for 1.5 mL Eppendorf tubes.
16. Microfuge.
17. Vortex mixer.
18. Thermal cycler.
19. P1000 pipette and tips.
20. P200 pipette and tips.
21. P100 pipette and tips.
22. P20 pipette and tips.
23. P10 pipette and tips.
24. P2 pipette and tips.
25. Ice bucket with ice.
26. Timer.
27. Agilent Bioanalyzer (optional).
28. Qubit fluorometer (or optional).
29. Eppendorf 5424 centrifuge (optional).

3 Method

3.1 DNA Extraction and Fragmentation

Extract DNA from your sample using a standardized protocol. The DNA was sheared using the Covaris g-TUBE fragmentation protocol with 2 μg input of DNA in 50 μL. Following which fragmentation was performed in the g-TUBE at 11,000 rpm with 30-s pulses.

Proceed with 48 μL of the fragmented DNA for the next step.

3.2 DNA Repair and End-Prep

1. Prepare the NEBNext FFPE DNA Repair Mix and NEBNext Ultra II End Repair/dA-tailing module reagents based on manufacturer's instructions:
   - All reagents must be thawed on ice.
   - Invert the reagent tubes to ensure they are mixed thoroughly.
   - Avoid vortexing the FFPE DNA Repair Mix or Ultra II End-Prep Enzyme Mix.
   - Spin down the tubes before opening them for the first time each day.
   - If there is any precipitate in the Ultra II End-Prep Buffer or FFPE DNA Repair Buffer, allow the mixture to come to room temperature, then pipette up and down to break up the precipitate, and vortex for 30 s to fully dissolve it. Thorough vortexing is important for proper mixing of the buffers.
   - The FFPE DNA Repair Buffer may appear yellow, which is normal and usable.
2. DNA prep in nuclease-free water.
   - Transfer 2 μg of the fragmented DNA into a 1.5 mL Eppendorf DNA LoBind tube and adjust the volume to 48 μL with nuclease-free water. Mix thoroughly by flicking the tube, and then briefly spin it in a microfuge.
3. PCR preparation:
   - In a PCR microtube, prepare the following reaction mixture:

Reagent	Volume (μL)
DNA from the previous step	48
NEBNext FFPE DNA repair buffer	3.5
NEBNext FFPE DNA repair mix	2
Ultra II end-prep Reaction Buffer	3.5
Ultra II end-prep enzyme mix	3
Total	**60**

4. Make sure the components are thoroughly mixed by pipetting and then spin down. Incubate using a thermal cycler at 20 °C for 5 min and at 65 °C for 5 min.
5. Resuspend the AMPure XP Beads (AXP) by vortexing and transfer the DNA sample to a clean 1.5 mL Eppendorf DNA LoBind tube.
6. Add 60 μL of resuspended AMPure XP Beads (AXP) to the end-prep reaction and mix by flicking the tube.
7. Incubate on a Hula mixer (rotator mixer) at room temperature for 5 min.
8. Prepare 500 μL of fresh 80% ethanol in nuclease-free water.
9. Spin down the sample, place it on a magnet until the supernatant is clear and colorless, then pipette off the supernatant while keeping the tube on the magnet.
10. Without disturbing the pellet, keep the tube on the magnet and wash the beads with 200 μL of freshly prepared 80% ethanol,
11. Discard the ethanol using a pipette and repeat this washing step.
12. Spin down the tube and return it to the magnet, then pipette off any remaining ethanol. Allow the beads to air-dry for about 30 s, but avoid drying them to the point where the pellet cracks.
13. Remove the tube from the magnet and resuspend the pellet in 61 μL of nuclease-free water. Incubate at room temperature for 2 min.
14. Pellet the beads on the magnet until the eluate is clear and colorless, for at least 1 min.
15. Transfer 61 μL of the clear eluate to a clean 1.5 mL Eppendorf DNA LoBind tube.

3.3 Adapter Ligation and Cleanup

1. Spin down the Ligation Adapter (LA) and Quick T4 Ligase and place them on ice.
2. Thaw the Ligation Buffer (LNB) at room temperature, and mix by pipetting (vortexing is ineffective due to its viscosity).
3. Place on ice immediately after mixing.
4. Thaw the Elution Buffer (EB) at room temperature, mix by vortexing, spin down, and then put on ice again.
5. Similarly, thaw the Long Fragment Buffer (LFB) at room temperature, mix by vortexing, spin down, and place on ice.
6. In a 1.5 mL Eppendorf DNA LoBind tube, prepare a reaction mixture:

Reagent	Volume (μL)
DNA sample from the previous **step 1**	60
Ligation Buffer (LNB)	25
NEBNext quick T4 DNA ligase	10
Ligation adapter (LA)	5
Total	**100**

7. Make sure the components are well-mixed by pipetting, then spin down. Incubate the reaction at room temperature for 10 min.
8. Resuspend the AMPure XP Beads (AXP) by vortexing and add 40 μL of the resuspended beads to the reaction and mix by flicking the tube.
9. Incubate on a Hula mixer (rotator mixer) at room temperature for 5 min. Spin down the sample and place it on a magnet to pellet the beads. Keep the tube on the magnet and pipette off the supernatant.
10. Wash the beads by adding 250 μL of Long Fragment Buffer (LFB). Flick the beads to resuspend, spin down, then return the tube to the magnetic rack and let the beads pellet. Remove and discard the supernatant. Repeat this washing step.
11. Spin down and place the tube back on the magnet. Pipette off any remaining supernatant and allow the beads to air-dry for about 30 s, but avoid drying to the point of cracking.
12. Remove the tube from the magnet and resuspend the pellet in 15 μL of Elution Buffer (EB). Spin down and incubate at room temperature for 10 min. For improved recovery of long fragments, consider incubating at 37 °C.
13. Pellet the beads on the magnet until the eluate is clear and colorless, for at least 1 min. Transfer 15 μL of the clear eluate containing the DNA library to a clean 1.5 mL Eppendorf DNA LoBind tube.
14. Transfer 150 ng of the final prepared library in 12 μL of Elution Buffer (EB).

3.4 Priming and Loading the SpotON Flow Cell

1. Thaw the Sequencing Buffer (SB), Library Beads (LIB), or Library Solution (LIS, if used), Flow Cell Tether (FCT), and one tube of Flow Cell Flush (FCF) at room temperature. Vortex to mix and then spin down.
2. Prepare the flow cell priming mix with Casein/polyvinyl alcohol by adding the reagents directly into the flow cell flush (FCF). Mix by inverting the tube and pipetting at room temperature.

Reagent	Volume per flow cell (μL)
Flow cell flush (FCF)	1170
Casein or polyvinyl alcohol	5
Flow cell tether (FCT)	30
Final total volume in flow cell flush(FCF) tube	**1205**

3. Open the lid of the MinION or GridION device and position the flow cell under the clip. Press down firmly on the flow cell to ensure proper thermal and electrical contact.
4. Turn the flow cell priming port cover clockwise to open the priming port.
5. Once the priming port is opened, inspect for a small air bubble under the cover. To remove any bubbles, follow these steps:
   - Set a P1000 pipette to 200 μL.
   - Insert the pipette tip into the priming port.
   - Turn the wheel until the dial reads 220–230 μL to withdraw 20–30 μL, or until you see a small volume of buffer in the pipette tip. (Ensure that there is a continuous flow of buffer from the priming port across the sensor array).
6. Load 800 μL of the priming mix into the flow cell through the priming port, taking care to avoid air bubbles. Allow it to sit for 5 min. Meanwhile, prepare the library for loading by following the steps below.
7. Completely combine the contents of the Library Beads (LIB) by using a pipette.
8. In a new tube, set up the library for loading with the following steps:

Reagent	Volume per flow cell (μL)
Sequencing Buffer (SB)	37.5
Library beads (LIB) mixed immediately before use, or library Solution (LIS), if using	25.5
DNA library	12
Total	**75**

9. Finish priming the flow cell by following these steps:
   - Carefully raise the SpotON sample port cover to access the SpotON sample port.

- Add 200 μL of the priming mix into the flow cell priming port (not the SpotON sample port), ensuring no air bubbles are introduced.

10. Gently mix the prepared library by pipetting up and down right before loading.
11. Add 75 μL of the prepared library to the flow cell through the SpotON sample port drop by drop, ensuring each drop fully enters the port before adding the next.
12. Carefully place the SpotON sample port cover back on, ensuring the bung fits into the SpotON port. Then, close the priming port and replace the lid of the MinION or GridION device.

3.5 Data Acquisition and Base Calling

MinKNOW software handles the control and data acquisition for the sequencing device. Ensure MinKNOW is installed on your computer. For additional setup instructions, refer to the MinKNOW protocol. Basecalling will be performed after sequencing in the Downstream Analysis section of the protocol.

1. Launch the MinKNOW software from the desktop shortcut and log in with your Community credentials. Then, choose your connected device.
2. To set up sequencing, select "Start sequencing."
3. Enter your experiment name and sample ID when prompted.
4. In the kit selection tab, choose the SQK-LSK114 kit from the dropdown menu.
5. In the run options tab, keep the run time and starting voltage at their default values. Under adaptive sampling, select "Enrich or deplete sequences." Expand the menu to upload the reference FASTA file and the bed file.
6. In the analysis tab, ensure that "Basecalling OFF" is selected.
7. Choose your desired location to save the data.
8. Review your sequencing setup and then click "Start."
9. For downstream analysis, make sure you have the latest version of Guppy installed. For performing basecalling and methylation calling, Remora is required. Prior expertise in programming and software operation is necessary.

3.6 Ending the Experiment

1. Once your sequencing experiment is finished, if you wish to reuse the flow cell, follow the Wash Kit instructions. Store the washed flow cell at 2–8 °C.
2. Follow the returns procedure to flush the flow cell before sending it back to Oxford Nanopore. Ensure all flow cells are flushed with deionized water.

4 Notes

1. Try an alternative extraction method to avoid contaminant carryover. Consider adding an extra SPRI cleanup step.
2. AMPure beads settle rapidly, so make sure they are thoroughly resuspended before adding them to the sample.
3. A lower AMPure beads-to-sample ratio results in stricter selection against short fragments. Always assess the input DNA length using an agarose gel (or other gel electrophoresis methods) and calculate the appropriate amount of AMPure beads to use accordingly.
4. Ensure you use the correct ethanol concentration since DNA will elute from the beads if ethanol is less than 70%.
5. After the Flow Cell Check, it is crucial to remove any air bubbles near the priming port before priming the flow cell. Air bubbles can travel to the nanopore array and cause irreversible damage.
6. If MinKNOW reports a lower number of pores at the start of sequencing compared to the number reported by the Flow Cell Check, stop the sequencing run, remove the flow cell, and reinsert it, ensuring it is securely seated and has reached the target temperature. If needed, try placing it in a different position on the device (GridION/PromethION).
7. During the Flow Cell Check, the pore count is measured using the QC DNA molecules in the flow cell storage buffer. At the beginning of sequencing, the library itself is used to estimate the number of active pores. This can lead to about 10% variability in pore count. A significantly lower pore count at the start of sequencing may indicate that contaminants in the library have damaged the membranes or blocked the pores. To improve the purity of the input material, consider using alternative DNA/RNA extraction or purification methods. Try an extraction method that minimizes contaminant carryover.
8. If there is a reduction in sequencing speed and q-score later in the run, add additional fuel to the flow cell as directed in the MinKNOW protocol. For future experiments, use a smaller amount of library when loading the flow cell.

References

1. Gouil Q, Keniry A (2019) Latest techniques to study DNA methylation. Essays Biochem 63(6): 639–648
2. Schmidt M, Van Bel M, Woloszynska M, Slabbinck B, Martens C, De Block M, Coppens F, Van Lijsebettens M (2017) Plant-RRBS, a bisulfite and next-generation sequencing-based methylome profiling method enriching for coverage of cytosine positions. BMC Plant Biol 17:1–14

Chapter 33

Reduced Representation Methylation Sequencing (RRMS) for Plant Viruses

Nulevino Iralu, Dasari Meghanath, Sumiah Wani, Sahar Saleem, and Aflaq Hamid

Abstract

Reduced Representation Methylation Sequencing (RRMS) is a powerful technique that has significant applications in the field of plant virology. RRMS is employed to analyze methylation changes, examine host response, study virus-induced epigenetic modifications, and understand gene regulation. RRMS is a valuable tool in plant virology, enhancing our understanding of the complex interactions between viruses and their host plants at the epigenetic level. RRMS can be tailored for detecting plant viruses by examining how viruses alter DNA methylation patterns in host plants This method offers insights into how viral infections influence plant epigenetics and can assist in detecting and tracking viral outbreaks in crops.

Key words RRMS, Methylation, DNA, Reaction, Patterns

1 Introduction

In RRMS, Oxford Nanopore's methylation detection is integrated with AS with the ability to target specific genome. This approach concentrates on areas rich in CpGs, including approximately 28,000 CpG islands, 50,600 shores, 42,700 shelves, and 21,600 promoter regions. One of its primary uses is in understanding the interactions between plant viruses and their host plants [1]. By comparing DNA methylation patterns in infected and uninfected plants, researchers can uncover how viruses influence the host's epigenome. This can reveal viral strategies for evading host defenses and highlight the ways in which viral infection alters the host's molecular landscape. In addition to studying host responses, RRBS can be applied to investigate the methylation status of viral genomes themselves. Understanding how viral DNA is methylated can provide insights into its stability and interaction with the host's cellular machinery, influencing viral replication and pathogenicity. The technique also allows for precise mapping of epigenetic

Aflaq Hamid et al. (eds.), *Detection of Plant Viruses: Advanced Techniques*, Springer Protocols Handbooks, https://doi.org/10.1007/978-1-0716-4390-7_33,

modifications at specific genomic loci. This high-resolution view helps scientists understand how these methylation changes impact gene expression, particularly regarding genes associated with plant defense mechanisms. By elucidating these regulatory networks, RRMS contributes to a deeper understanding of plant immunity and the genetic basis of viral resistance. RRMS facilitates evolutionary studies by allowing comparisons of methylation patterns across different plant species or varieties in response to similar viral stresses. This can reveal evolutionary adaptations in plant virology, shedding light on how different plants cope with viral infections over time [2]. RRMS enables the characterization of the host plant's response to viral infection. By examining changes in DNA methylation, researchers can identify specific pathways involved in resistance or susceptibility to various viral pathogens. This knowledge is crucial for developing strategies to enhance plant resilience against viral threats, thereby contributing to improved crop security and yield [3]. Primary focus revolves not only on DNA methylation but may target a broader range of methylation marks, not just CpG sites. It often uses different techniques to enrich for specific regions or modifications [4].

2 Materials

1. Genomic DNA: High-quality DNA extracted from your sample.
2. Restriction Enzymes: To digest the DNA and target specific regions of the genome, often focusing on regions enriched in CpG sites.
3. Methylation-Specific Enrichment Kits: Reagents and protocols for capturing and enriching methylated regions of the genome.
4. Library Preparation Kit: Includes adapters and reagents necessary to prepare DNA libraries for sequencing.
5. PCR Reagents: For amplifying the prepared libraries if required.
6. Sequencing Platform: A sequencer compatible with the prepared libraries (e.g., Illumina sequencer).
7. Quality Control Reagents: For assessing the quality of DNA and library preparation, such as gel electrophoresis or bioanalyzer tools.
8. Buffers and Solutions: Required for DNA digestion, enrichment, and library preparation.
9. Clean-Up Kit: For purifying the DNA after enrichment and library preparation steps.

3 Method

3.1 Sample Preparation

1. Tissue Collection: Gather plant tissues of interest, ensuring they are either fresh or well-preserved to prevent degradation.
2. DNA Extraction: Isolate genomic DNA from the collected tissues using a DNA extraction kit designed for plant samples, and adhere to the manufacturer's guidelines to achieve the best yield and purity (or refer to Chapter 13).

3.2 DNA Digestion

1. Restriction Enzyme Digestion: Digest the extracted DNA with a restriction enzyme that cuts outside of CpG sites (e.g., MspI or HpaII). This will create a reduced representation of the genome focusing on CpG-rich regions.
2. Prepare Digestion Reaction: Mix DNA with the restriction enzyme buffer and the enzyme itself according to the standardized guidelines.
3. Incubation: Incubate the reaction at the optimal temperature for the enzyme (usually 37 °C) for a specified time (often overnight).

3.3 Bisulfite Treatment

1. Bisulfite Conversion: Subject the digested DNA to bisulfite treatment, which converts unmethylated cytosines into uracils while preserving methylated cytosines.
2. Prepare Bisulfite Solution: Use a bisulfite conversion kit, following the manufacturer's protocol for mixing and applying the bisulfite reagent.
3. Conversion Process: Perform the bisulfite conversion under the recommended conditions, often involving multiple cycles of heating and cooling.

3.4 Library Preparation

1. PCR Amplification: Amplify the bisulfite-treated DNA through PCR to enrich the DNA fragments of interest.
2. Prepare PCR Reaction Mix: Combine bisulfite-treated DNA with PCR primers specific to the regions of interest, DNA polymerase, and other PCR reagents (refer Chapter 20).
3. Cycling Conditions: Follow the standardized cycling conditions for the PCR amplification specific to the primers being used.
4. Library Construction: Construct sequencing libraries from the PCR products.
5. Fragmentation and Adapter Ligation: Fragment the PCR products and ligate sequencing adapters to the ends of the fragments.
6. Purification: Purify the adapter-ligated fragments to remove any contaminants.

3.5 Sequencing

1. Prepare Sequencing Library: Quantify and assess the quality of the sequencing library.
2. Sequencing: Sequence the libraries using a high-throughput sequencing platform (e.g., Illumina). Choose the appropriate read length and depth based on the research needs.

3.6 Data Analysis

1. Preprocessing: Clean and trim raw sequencing reads to remove low-quality sequences and adapters.
2. Alignment: Align the reads to a reference genome using a bisulfite-specific aligner (e.g., Bismark).
3. Methylation Analysis: Analyze the aligned data to identify and quantify DNA methylation levels at CpG sites.
4. Comparison and Interpretation: Compare methylation patterns between samples or conditions to identify significant changes and correlations.

3.7 Validation (Optional)

1. Experimental Validation: Validate key findings using alternative methods such as bisulfite sequencing of specific regions, qPCR, or methylation-specific PCR.

3.8 Documentation and Reporting

1. Record Keeping: Document all experimental conditions, observations, and results meticulously.
2. Report Findings: Prepare a detailed report of the findings, including methods, data, analysis, and interpretations.

4 Notes

1. Sample Handling: Handle plant tissues and DNA samples carefully to prevent contamination and degradation.
2. Enzyme Activity: Use restriction enzymes and bisulfite reagents as recommended to prevent incomplete digestion or conversion.
3. Bisulfite Treatment: Perform bisulfite treatment under controlled conditions to ensure complete conversion of unmethylated cytosines.
4. PCR Conditions: Optimize PCR conditions to prevent non-specific amplification and ensure accurate representation of the methylation status.
5. Library Quality: Assess the quality of sequencing libraries before sequencing to avoid errors and ensure reliable results.
6. Data Management: Use appropriate software and methods for data analysis to accurately interpret methylation patterns and minimize errors.

References

1. Alvarado Marchena LF (2022) Insights into the molecular mechanisms of the n6-methyladenosine (m6a) methylation machinery in the regulation of the infection cycle of rna plant viruses. Doctoral dissertation, Universitat Politècnica de València
2. Glushkevich A, Spechenkova N, Fesenko I, Knyazev A, Samarskaya V, Kalinina NO, Taliansky M, Love AJ (2022) Transcriptomic reprogramming, alternative splicing and RNA methylation in potato (Solanum tuberosum L.) plants in response to potato virus Y infection. Plan Theory 11(5):635
3. Gouil Q, Keniry A (2019) Latest techniques to study DNA methylation. Essays Biochem 63(6): 639–648
4. Cristalli C, Scotlandi K (2024) Targeting DNA methylation machinery in pediatric solid tumors. Cells 13(14):120

Chapter 34

Reduced Representation Bisulfite Sequencing (RRBS) for Plant Viruses

Nulevino Iralu, Dasari Meghanath, Sumiah Wani, Sahar Saleem, and Aflaq Hamid

Abstract

In plant virology, RRBS (Reduced Representation Bisulfite Sequencing) is employed to assess changes in DNA methylation related to viral infections and plant reactions, understand host-virus dynamics, detect epigenetic changes, and investigate gene regulation. RRBS can be applied to plant virus detection by analyzing how viral infections alter DNA methylation patterns in host plants. This approach helps detect viruses by revealing how they disrupt normal methylation patterns in plants. By elucidating the complex interactions between viruses and their host plants, RRBS can help inform strategies for enhancing plant health and resilience in the face of viral challenges.

Key words Bisulfite, Sequencing, Methylation, DNA

1 Introduction

Reduced Representation Bisulfite Sequencing (RRBS) is a specialized technique used to study DNA methylation by focusing on specific regions of the genome [1]. RRBS is an innovative approach that provides insights into DNA methylation patterns and their role in plant virology. This technique is particularly valuable for understanding the epigenetic changes that occur in plants upon viral infection. By focusing on regions of the genome that are enriched for regulatory elements, RRBS allows for a detailed examination of how viruses impact host DNA methylation [2]. One of the primary applications of RRBS in plant virology is investigating the interplay between viral infection and the host's epigenetic landscape. When a plant is infected by a virus, it can undergo significant changes in DNA methylation, which may affect gene expression related to defense mechanisms. RRBS enables researchers to compare methylation patterns before and after infection, helping to identify specific genes or pathways that are modulated

Aflaq Hamid et al. (eds.), *Detection of Plant Viruses: Advanced Techniques*, Springer Protocols Handbooks, https://doi.org/10.1007/978-1-0716-4390-7_34,

by the virus. This understanding can reveal how plants utilize epigenetic regulation to respond to viral threats [3]. Primary focus of RRBS is on DNA methylation analysis, specifically on CpG-rich regions [4]. RRBS also provides a platform for studying the methylation status of viral genomes. Understanding how viral DNA is methylated can inform researchers about its stability, replication, and interactions with host factors. This can be critical for unraveling the life cycle of the virus and its impact on the host plant [5]. Additionally, RRBS can shed light on the strategies employed by viruses to manipulate host gene expression [6]. Some viruses have been shown to induce changes in the host's DNA methylation to suppress antiviral defense responses. By mapping these alterations with RRBS, researchers can gain insights into the molecular mechanisms underlying viral pathogenesis and identify potential targets for genetic engineering or breeding strategies aimed at enhancing resistance in crops [7]. Finally, RRBS can facilitate comparative studies across different plant species or cultivars. By analyzing how various plants respond epigenetically to the same viral infection, researchers can uncover evolutionary adaptations and resilience mechanisms [8]. This comparative approach can ultimately contribute to the development of more resistant plant varieties through informed breeding practices. Reduced Representation Bisulfite Sequencing is a powerful tool in plant virology, offering a comprehensive understanding of the epigenetic dynamics that occur during viral infections.

2 Materials

1. High-quality DNA extracted from your sample.
2. Restriction Enzymes to digest the DNA and generate a reduced representation of the genome, concentrating on CpG-rich regions.
3. Bisulfite Conversion Kit: To convert unmethylated cytosines to uracil while preserving methylated cytosines, thereby preparing the DNA for sequencing.
4. Library Preparation Kit: Includes adapters and reagents for preparing sequencing libraries from the bisulfite-converted DNA.
5. PCR Reagents: For amplifying the library if necessary.
6. Sequencing Platform: A sequencer compatible with the prepared library (e.g., Illumina sequencer).
7. Quality Control Reagents: For checking DNA quality and library preparation, such as gel electrophoresis or bioanalyzer reagents.

8. Buffers and Solutions: For DNA digestion, bisulfite conversion, and library preparation steps.
9. Clean-Up Kits: For purifying the DNA after bisulfite treatment and during library preparation.

3 Method

3.1 Sample Preparation

1. Tissue Collection: Collect plant or animal tissues and ensure they are fresh or properly preserved to avoid degradation.
2. DNA Extraction: Extract genomic DNA from the collected tissues using a DNA extraction kit suitable for the sample type. Follow the manufacturer's protocol for optimal yield and purity.

3.2 Restriction Enzyme Digestion

1. Selection of Enzyme: Choose restriction enzymes that cut outside of CpG sites, such as MspI or HpaII. These enzymes create a reduced representation of the genome, focusing on CpG-rich regions.
2. Digestion Reaction.
3. Prepare a digestion mixture containing the genomic DNA, enzyme buffer, and the restriction enzyme.
4. Incubate the reaction at the optimal temperature for the enzyme (typically 37 °C) for a specified time (often overnight).

3.3 Bisulfite Treatment

1. Bisulfite Conversion:
2. Treat the digested DNA with bisulfite to convert unmethylated cytosines into uracils while leaving methylated cytosines unchanged.
3. Use a bisulfite conversion kit and follow the manufacturer's instructions for preparing and applying the bisulfite reagent.
4. The process typically involves a series of heating and cooling steps to ensure complete conversion.

3.4 Library Preparation

1. PCR Amplification:
2. Amplify the bisulfite-treated DNA using PCR. This step enriches the fragments of interest.
3. Prepare the PCR reaction mix with bisulfite-treated DNA, PCR primers specific to the target regions, DNA polymerase, and other PCR reagents.
4. Perform PCR amplification following the recommended cycling conditions.
5. Library Construction:

6. Fragment the PCR products (if necessary) and ligate sequencing adapters to the ends of the fragments.
7. Purify the adapter-ligated DNA fragments to remove contaminants and unwanted products.

3.5 Sequencing

1. Library Quality Control: Assess the quality and concentration of the sequencing libraries using methods such as a Bioanalyzer or qPCR.
2. Sequencing: Sequence the prepared libraries using a high-throughput sequencing platform (e.g., Illumina). Select appropriate read length and depth based on the research goals.

3.6 Data Analysis

1. Preprocessing: Clean and trim raw sequencing reads to remove low-quality sequences and adapter remnants.
2. Alignment: Align the cleaned reads to a reference genome using a bisulfite-specific aligner, such as Bismark or BS-Seeker.

3.7 Methylation Analysis

1. Analyze the alignment data to determine the methylation status of CpG sites.
2. Quantify the methylation levels and identify differentially methylated regions.
3. Interpretation: Compare methylation patterns between different conditions or samples to draw conclusions about epigenetic changes.

3.8 Validation (Optional)

1. Experimental Validation: Validate key findings using alternative methods such as bisulfite pyrosequencing or methylation-specific PCR.

3.9 Documentation and Reporting

1. Record Keeping: Document all experimental conditions, protocols, and results meticulously.
2. Report Findings: Prepare a detailed report summarizing methods, data, analysis, and interpretations.

4 Notes

1. Sample Quality: Plant tissues must be handled with caution to avoid any degradation or contamination. Ensure that the DNA is high-quality and free from contaminants that could affect enzyme performance.
2. Enzyme Handling: Follow the manufacturer's instructions for using restriction enzymes and bisulfite reagents to ensure complete digestion and conversion. Verify the enzyme activity before use.

3. Bisulfite Treatment: Perform bisulfite conversion under recommended conditions to ensure complete and consistent conversion of unmethylated cytosines to uracils. Minimize exposure to air and light during this process to reduce degradation.
4. PCR Optimization: Optimize PCR conditions to avoid nonspecific amplification and ensure efficient amplification of bisulfite-treated DNA. Use controls to monitor for contamination and amplification artifacts.
5. Library Quality: Assess the quality and concentration of sequencing libraries before sequencing to ensure they meet the required specifications. Incomplete or poor-quality libraries can lead to unreliable data.
6. Data Handling: Use appropriate software for aligning bisulfite-treated sequences and analyzing methylation data. Ensure proper handling and storage of sequencing data to prevent loss or corruption.
7. Contamination Prevention: Keep a clean working environment to prevent cross-contamination between samples. Use separate equipment and reagents for each stage of the protocol.

References

1. Xi Y, Bock C, Müller F, Sun D, Meissner A, Li W (2012) RRBSMAP: a fast, accurate and user-friendly alignment tool for reduced representation bisulfite sequencing. Bioinformatics 28(3): 430–432
2. Sun Y, Fan M, He Y (2019) DNA methylation analysis of the Citrullus lanatus response to cucumber green mottle mosaic virus infection by whole-genome bisulfite sequencing. Genes 10(5):344
3. Wang B, Yang X, Wang Y, Xie Y, Zhou X (2018) Tomato yellow leaf curl virus V2 interacts with host histone deacetylase 6 to suppress methylation-mediated transcriptional gene silencing in plants. J Virol 92(18):10–1128
4. Klughammer J, Datlinger P, Printz D, Sheffield NC, Farlik M, Hadler J, Fritsch G, Bock C (2015) Differential DNA methylation analysis without a reference genome. Cell Rep 13(11): 2621–2633
5. Meissner A, Gnirke A, Bell GW, Ramsahoye B, Lander ES, Jaenisch R (2005) Reduced representation bisulfite sequencing for comparative high-resolution DNA methylation analysis. Nucleic Acids Res 33(18):5868–5877
6. Wang C, Wang C, Xu W, Zou J, Qiu Y, Kong J, Yang Y, Zhang B, Zhu S (2018) Epigenetic changes in the regulation of Nicotiana tabacum response to cucumber mosaic virus infection and symptom recovery through single-base resolution methylomes. Viruses 10(8):402
7. Saeed F, Chaudhry UK, Bakhsh A, Raza A, Saeed Y, Bohra A, Varshney RK (2022) Moving beyond DNA sequence to improve plant stress responses. Front Genet 13:874648
8. Schmidt M, Van Bel M, Woloszynska M, Slabbinck B, Martens C, De Block M, Coppens F, Van Lijsebettens M (2017) Plant-RRBS, a bisulfite and next-generation sequencing-based methylome profiling method enriching for coverage of cytosine positions. BMC Plant Biol 17:1–14

Chapter 35

Remote Sensing for Virus Detection

Insha Yaqoob, Nulevino Iralu, Sumiah Wani, and Aflaq Hamid

Abstract

The worldwide demand for agricultural products exceeds the supply, and plant viral diseases result in productivity and economic losses to agriculture, necessitating accurate detection for effective control. Lab-based molecular testing provides reliable and accurate diagnostics, but these tests are expensive, time-consuming, and labor-intensive, especially at the field-scale with a large number of samples. The implementation of noncontact, highly efficient, and affordable methods for detecting and monitoring plant diseases and pests over vast areas could greatly facilitate plant protection. In this respect, different forms of remote sensing methods have been introduced for detecting and monitoring plant diseases and pests in many ways. The use of innovative technologies like "Remote sensing" holds promise in agricultural crop production including crop protection. It is a rapid, noninvasive, and efficient technique, which can acquire and analyze spectral properties of earth surfaces from various distances, ranging from satellites to ground-based platforms. This chapter provides an overview of the science behind the remote sensing, general material, and methodology involved and the various remote sensing systems available for monitoring plant viral diseases including RBG cameras, Multispectral imagery, Hyperspectral sensing, Chlorophyll fluorescence (Chl-Fl) and Infrared (IR) thermal sensors and SAR and Lidar systems. Finally, the current challenges and considerations for detecting plant viruses are discussed.

Key words Remote sensing, RBG cameras, Multispectral imagery, Hyperspectral sensing, Chlorophyll fluorescence (Chl-Fl) and Infrared (IR) thermal sensors, Synthetic aperture radar (SAR), Light Detection and Ranging (Lidar)

1 Introduction

Viral infections hinder plant growth, diminish yields, and compromise fruit quality, resulting in substantial economic repercussions [1] Understanding the location, spread, and severity of these threats is crucial for effective plant protection. Traditional methods of field inspection are time-consuming, inconsistent, and inefficient. Remote sensing technology offers a promising alternative for large-scale monitoring of plant health and identifying disease or pest outbreaks [2].

The "American Society for Photogrammetry and Remote Sensing (ASPRS)" defined photogrammetry and remote sensing as,

Aflaq Hamid et al. (eds.), *Detection of Plant Viruses: Advanced Techniques*, Springer Protocols Handbooks,
https://doi.org/10.1007/978-1-0716-4390-7_35,

"the art, science and technology of obtaining reliable information about physical objects and the environment, through the process of recording, measuring and interpreting imagery and digital representations of energy patterns derived from non-contact sensor systems." It is also defined as the science of acquiring information about objects at or near the surface of the Earth and atmosphere based on radiation reflected or emitted from those objects without having direct contact with it. Remote sensing provides a rapid and noninvasive method for analyzing plant health from various distances. Application of remote sensing data in epidemiology involves retrieving environmental variables such as land cover, temperature, humidity or vapor pressure, and precipitation that influence the ecosystem [3]. With a rapid development of computer science and sensing technology, a wide spectrum of RS data has been used to detect diseases and pests. By capturing and examining the light reflected by plants, we can identify signs of diseases or pests. Techniques like multispectral and hyperspectral imaging, as well as thermal and fluorescence measurements, are particularly useful for detecting plant stress. These technologies offer the potential to revolutionize agriculture by enabling early disease detection, reducing pesticide use, and optimizing crop management.

For the last few decades, studies on sensing systems, feature extraction, and algorithms have been carried out at multiple scales, revealing new possibilities in remotely monitoring plant diseases and pests [2, 4–8]. In this process, connections between RS methodologies and plant pathology theories have been strengthened thus improving our understanding of the agriculture system in many aspects.

1.1 Science Behind the Remote Sensing Technology

1.1.1 Reflection Pattern in Healthy Vegetation

Vegetation reflectance is categorized into three primary spectral domains, i.e., visible, near-infrared, and middle-infrared, each forming the basis for distinct remote sensing techniques. Sahoo et al. [9] elucidated that a plant's spectral signature is intricately linked to its biophysical and biochemical properties, including leaf area, biomass, pigments, moisture content, and cellular structure. Chlorophyll a and b, carotenoids, xanthophylls, and polyphenols are the main light absorbing pigments in the visible spectrum (0.4–0.7 μm) creating characteristic absorption bands and a prominent reflectance peak around 0.55 μm. In contrast, the near-infrared (NIR) region (0.7–1.3 μm) exhibits high reflectance due to internal scattering within leaves. The mid-infrared (MIR) or shortwave- infrared (SWIR) domain (1.3–2.5 μm) is strongly affected by water content and other foliar constituents, with prominent absorption bands at specific wavelengths [9].

1.1.2 Reflection Pattern in Sick and Dead Leaves

Plant stress or disease alters the plant's biochemical and physical makeup, causing changes in how it reflects light. A stressed plant often produces less chlorophyll, leading to reduced absorption in the blue and red wavelengths. Consequently, these wavelengths are reflected more strongly, giving the plant a yellowish or brownish appearance. Additionally, stressed or diseased plants absorb more near-infrared (NIR) light in the mesophyll cells, resulting in darker tones in images.

2 Materials

1. Sensors: Sensors are crucial in detecting plant stress caused by viruses (*see* **Note 1**). Common types include:
   (a) Multispectral and Hyperspectral Sensors: These sensors measure the reflectance of an object, such as plant leaves, and identify any changes, such as those caused by a plant pathogen, as any disturbance to the leaf region would alter the reflectance and direction of the light [10]. Multispectral sensors capture data from a few specific bands (e.g., visible and near-infrared) and are useful for identifying basic plant stress symptoms whereas hyperspectral sensors capture data from hundreds of narrow, continuous spectral bands, allowing detailed analysis of plant physiology. Hyperspectral sensors provide more detailed spectral information compared to multispectral sensors [11].
   (b) Thermal Infrared Sensors: These sensors measure the thermal emission from plants, which can be indicative of stress caused by viral infections.
   (c) RGB Cameras: Standard visible light cameras can also be used to detect visual symptoms of viral infections, although they are less effective compared to multispectral and hyperspectral sensors.
   (d) Fluorescence Sensors: These sensors detect florescence emission from plants under ultraviolet light, which may be indicative of stress or viral infection.
2. Remote Sensing Platforms: Remote sensing systems can be mounted on various platforms, depending on the scale and resolution needed (*see* **Note 2**).

   Satellites (e.g., Landsat, Sentinel, MODIS) provide large-scale coverage and are useful for monitoring extensive agricultural areas and tracking the spread of viral infections over larger areas.
3. Unmanned Aerial Vehicles (UAVs or drones) offer high-resolution imagery and can be used for more localized and detailed monitoring.

   Drones can be equipped with various types of sensors and cameras.

Ground-Based Systems: Mobile platforms like all-terrain vehicles equipped with sensors can be used for close-range monitoring.

4. Data Processing Tools
   - (a) Image Preprocessing: This involves using softwares that correct for atmospheric interference, sensor noise, and other distortions to enhance the quality of the captured images. ENVI, ArcGIS, QGIS, Python libraries (e.g., OpenCV, Scikit-image).
   - (b) Vegetation Indices: Indices such as NDVI (Normalized Difference Vegetation Index) Chlorophyll Index, Water Index, etc., are used to assess plant health and detect stress. Changes in these indices can indicate the presence of viral infections [12].
   - (c) Machine Learning and AI: Supervised and unsupervised machine learning algorithms such as random forests, support vector machines, and neural networks are used to classify and detect viral infections from remote sensing data.
   - (d) Geographic Information Systems (GIS): GIS platforms are used for spatial analysis and visualization of data collected from remote sensing.

3 Methods

1. Image Acquisition:

   Choose a suitable remote sensing platform and sensor based on spatial and spectral resolution requirements. Capture images of the target crop area at regular intervals (e.g., daily, weekly) (*see* **Note 3**).

   Consider factors like weather conditions, sun angle, and atmospheric conditions that can affect image quality (*see* **Note 4**).

2. Pre-processing:

   Correct for atmospheric effects (e.g., haze, clouds) using atmospheric correction algorithms.

   Geometrically correct images to align them with a common coordinate system.

   Radiometrically calibrate images to ensure accurate reflectance values (*see* **Note 5**).

3. Feature Extraction:

   Calculate spectral indices (e.g., NDVI, EVI, SAVI) to enhance vegetation information. The relationship between spectral reflectance and vegetation properties is well established and is to a large degree based on chlorophyll a and b spectral properties (*see* **Note 6**).

Derive textural features to capture spatial patterns associated with disease symptoms.

Extract phenological metrics (e.g., time of green-up, senescence) to monitor crop development.

4. Classification and Detection:

   Use supervised or unsupervised classification techniques to identify diseased and healthy plants.

   Implement machine learning algorithms (e.g., support vector machines, random forests) for accurate classification.

   Develop disease detection models based on spectral, textural, and phenological features.

5. Validation:

   Ground truthing is a critical part of the remote sensing methodology. It involves collecting data directly from the field to confirm the accuracy of remote sensing observations.

   Compare classification results with ground truth data to assess accuracy (*see* **Note 7**).

   Calculate error matrices and overall accuracy to evaluate model performance.

6. Monitoring and Early Warning:

   Integrate remote sensing data with ground-based observations for early disease detection.

   Develop early warning systems to alert farmers and authorities about disease outbreaks.

3.1 The Remote Sensing Systems Available for Monitoring Plant Diseases and Pests

The objectivity of field-based visual assessment is increased by the sensor technology, which offers quick and nondestructive alternatives to molecular approaches for plant disease diagnosis [13, 14]. Optical sensors measure the frequency and intensity of light, both of which can be interpreted as meaningful information using multivariate statistical techniques. Various parts of the electromagnetic spectrum, ranging from ultraviolet (UV) to long-wave infrared (LWI), can be measured by the various sensors. Using optical sensing technologies, subtle phenotype changes caused by the disease are detectable [12]. Optical sensors can be employed handheld, in close proximity to ground vehicles, and directly in contact with leaves at the finest scale. Certain non-imaging sensors, such as spectroradiometers, GreenSeeker thermometers, laser thermometers, and chlorophyll fluorimeters, are typically employed in close proximity or direct contact with leaves [15]. Imaging sensors can be used to support regional disease control either proximally, like when installed on tractors, or remotely, such when employed on airborne platforms like unmanned aerial vehicles (UAV), aircraft, and satellites, via increasing the sensing distance [16].

In general, spatial resolution decreases with increasing sensing distance. Although they have the lowest spatial resolution, satellite photos offer the widest coverage of the land. In a moderate region, manned aircraft can obtain photographs with a higher spatial resolution than those from satellites. Drones or UAVs equipped with lightweight optical sensors have the ability to fly as low as a few meters above the ground [17–19], potentially delivering images with millimeter spatial resolution for the detection of plant viral diseases. There is a wide variety of sensing technologies that can be used for plant viral disease detection. We find the RGB (red, green, and blue), multispectral and hyperspectral systems arranged in increasing order of spectral detail. Because RGB systems are as sensitive as the human eye, they provide easily comprehended visuals [20].

3.2 RBG Cameras

The raw data commonly used for disease detection is RGB images, which are generally acquired by digital camera. Several disease detection models, which combine RGB images with machine learning, were proposed in recent years [21]. Due to their ease of use and public availability, modern RGB cameras can provide access to a multitude of datasets for the identification of plant diseases. Recent studies employing computer vision algorithms have used community-shared libraries of photos contaminated with plant diseases to identify plant diseases [22, 23]. In addition, low altitude, high-resolution aerial photos from UAV have produced a high-popularity in-field plant disease detection [14, 24, 25].

3.3 Multispectral Imagery

Various spectral wavelength regions throughout the electromagnetic spectrum can be measured by multispectral sensors; the selection of these regions or bands is contingent upon the intended use [26]. Multispectral cameras have been commonly used in remote sensing to explore land use, characterize vegetation, and monitor the environment and urban structures [27]. The vegetation characteristics of light absorption and reflection at various wavelengths are taken into consideration while choosing the spectral bands in multispectral cameras for agricultural applications. Typically, RGB, combined with the unique vegetative reflectance regions, red edge (690–740 nm) and near-infrared (NIR) (700–1300 (tel:700–1300) nm) bands, are used in multispectral sensors [28, 29]. This information enables the computation of specific vegetation indices (VIs) that can be used to evaluate different characteristics of the vegetation. For example, by contrasting the absorption and reflectance in red and NIR spectral regions, the well-known normalized difference vegetation index (NDVI) can be calculated [30, 31]. Multispectral photography is an easy, affordable, and accessible way to monitor plants. Spectral analysis of individual pixels is the emphasis of its established methodology, in contrast to object-based detection methods like as CV, which use complete images or subsets of images. The spectrum characteristics of chlorophyll a and b form a major basis for the well- established link between spectral reflectance and vegetative features. For chlorophyll a and b, the absorption bands are around 440 and 650 nm, respectively, with strong reflectance in the green and near-infrared spectral ranges [32, 33].

There are more than 150 published vegetation indices (VI), and different relationships with plant cell structure, biochemistry, physiology, and stress have been established [34, 35]. Commonly used VIs for plant stress studies include NDVI, Chlorophyll Index, Water Index, and Red-edge Vegetation Stress Index [36]. Multispectral sensors can focus on these important spectral bands and produce a variety of indices that have been examined for disease detection.

3.4 Hyperspectral Sensing

Unlike multispectral sensors, which only record a few distinct broad bands, hyperspectral sensors record hundreds of contiguous small bands (2–20 nm) over a wide range of spectra (UV, visible, near-infrared (VNIR), to short-wave infrared (SWIR)) [37]. The highest spectral detail is obtained by sensing single point spectroradiometers rather than imaging, for example, ASD FieldSpec 4 and Ocean Optics USB4000. The majority of the time, these sensors are employed in close proximity or directly in contact with the plant. Numerous investigations have revealed the relationship between plant viral infections and spectral reflection signals, indicating the possibility of this approach for early or asymptomatic stage identification [38–42]. A hyperspectral imaging system provides both spatial and spectral information to produce a 3-dimensional (3D) data cube [43]. The X-Y dimension expresses the spatial position information of the image. The third dimension (λ) is the spectral/wavelength dimension, which is composed of several bands in the spectral space. By making a multidimensional section of the spectral image cube, different types of spectral features can be obtained, such as the spectral characteristics at any point pixel, the spectral change of a spectral interval on an arbitrary spatial profile, or the spatial image of any band in the spectral dimension, etc. With the above information, we can not only identify the object according to the image feature in the spatial section but also analyze the spectral feature in the spectral dimension. As a result, it will benefit the recognition of the types, components, and contents of the substances. Hyperspectral imaging data present a potentially significant advantage in plant disease studies at broader scales and finds application in the identification of viral infections before visible symptoms appear and distinguishing between different types of plant stress (e.g., viral vs. bacterial infections).

Hyperspectral imaging data present a potentially significant advantage in plant disease studies at broader scales.

3.5 Chlorophyll Fluorescence (Chl-Fl) and Infrared (IR) Thermal Sensors

Thermal infrared and fluorescence RS systems are able to track the respiration and photosynthetic processes of plants, which allow pre-symptomatic monitoring of plant diseases and pests.

An essential metric for plant health and stress expression is Chl-Fl [44, 45]. Numerous research studies have shown the possibility of employing Chl-Fl to identify plant viral infection at an early illness stage and have employed Chl-Fl as the laboratory's analytical tool for plant virus infection [46–51]. The passive method that uses solar radiation to measure fluorescence rate (known as solar-induced Chlorophyll fluorescence) remotely for vegetation stress has also been attempted for plant stress detection [52–58].

To obtain the fluorescence response of plants, laser-induced fluorescence (LIF) is the most promising technique. Some commercialized fluorometers, such as portable chlorophyll fluorometer PAM series (i.e., PAM2100, PAM2500, IMAGINE-PAM, etc. from Heinz Walz GmbH), apply saturation pulse technique and induced LED light sources at fixed wavelength (i.e., 655 and 735 nm) to generate some important fluorescence indicators (e.g., the maximal quantum yield (Fv/Fm), non-photochemical fluorescence quenching (NPQ), the operating efficiency of PSII (ΦPSII), etc.), and to record some kinetics curves and parameters [59]. In addition, by coupling laser emitting light source (e.g., He–Ne light source) with an analytical spectrometer (e.g., FieldSpec TM UV/ VNIR, ASD Inc., Boulder, CO., USA), the continuous fluorescence spectra can be measured, allowing the extraction of fluorescence features at a continuous wavelength [60].

In addition to the fluorescence response, the infection of diseases and pests may sometimes change the respiration and evapotranspiration rates of plants, which can lead to noticeable shifts of thermal emission levels [61]. Thermal cameras detect heat signatures emitted by plants, which can change due to viral infections affecting plant metabolism and water relations. Infrared thermography is used to measure leaf and canopy temperatures, which can indicate plant stress caused by viral infections. The thermal sensor is predominantly used in precision agriculture to detect and monitor crop biotic and abiotic stress [18, 62], and spatial and temporal thermography patterns have shown potential for the early detection of viral disease.

3.6 Synthetic Aperture Radar (SAR) and Light Detection and Ranging Equipment (Lidar) Systems

SAR and Lidar can be used to delineate plant characteristics and their environments, which can potentially be used in the monitoring of diseases and pests [59]. As an active way of remote sensing, the cloud penetrating capability of SAR enables monitoring in all-weather condition [63]. With the development toward full polarimetric and high resolution, SAR has been used in retrieving plant water content, soil characteristics, and some structural parameters of plant canopy, especially for some tall plants (e.g., trees, corn, sugarcane, etc.) [64]. These observations can potentially be used for delineating the habitat characteristics of plant viral diseases.

While primarily used for structural analysis, LiDAR can also detect changes in plant architecture caused by viral infections. The Lidar system is capable of obtaining detailed canopy morphology information. The point cloud data for plant canopy, collected by Lidar sensors, enables the reconstruction of 3D canopy structure, which can indicate damages or biomass loss due to diseases [59]. LiDAR systems emit laser pulses and measure the time it takes for the reflections to return, creating detailed 3D models of plant structure. Some plant diseases can alter plants' canopy morphology significantly [65] and could be potentially detected by Lidar sensors. Thus, it might be possible to discriminate their occurrence and impact looking at temporal series of Lidar data over a crop season.

3.7 Challenges and Considerations

Plant viral disease detection has come a long way, but there are still many obstacles that new technology can help with. Advances in molecular diagnostic techniques have led to a greater understanding of plant viruses; however, because of the intricate interactions between viruses, host genomes, and environmental factors, as well as the fact that symptoms are frequently not uniformly distributed throughout the plant, determining the plant phenotypes resulting from virus infections remains difficult [14]. The effects of the viral disease vary; some people experience no symptoms at all, while others see a sharp drop in plant life [56]. It's not always the case that viruses infect plants and cause disease. Evidence of a certain degree of virus tolerance has been established by the recovery of some infected plants even when the virus is still present in the host, which has not received much attention to date. All these complexities associated with viruses in plants make disease detection a challenging task.

Multiple interactions between the host, pathogen, and environment can cause a wide range of symptoms in viral diseases. Plants that are coinfected with many pathogens may have more difficulty detecting viruses. The signs of viral diseases can also be confused with those of other pathogens, such as bacteria, fungi, nematodes, and viroids; they can also be misinterpreted as the result of abiotic stressors such as water stress or nutrient shortages (such as potassium or phosphorus). It's also necessary to take into account environmental influences including soil type, air temperature, and edge effects. Herbicide injuries and other mechanical and chemical damage may induce stress responses in plants that are comparable to those caused by viral diseases. The effective detection of viruses may be hampered by these intricate combinations of variables. Therefore, it's necessary to have a thorough awareness of the plants' state and to monitor them continuously over time to improve the accuracy of viral disease detection.

Remote sensing technology relies heavily on the availability of precise ground truth data from various dates, disease severities, and geographical locations in order to build a reliable disease prediction model. Large-scale data acquisition is still difficult and requires resources, labor, and time. A straightforward technique such as visual evaluation can collect enough ground truth information to train a model. Its dependability is inadequate, nevertheless, as some infected plants show no outward signs of infection, which could result in misleading negative results. Although lab-based testing techniques are reliable and necessary for ground truthing, their high expense restricts the amount of ground truth data available, which could weaken the model's resilience. However, none of the diagnostic techniques can provide perfect accuracy.

The other noteworthy issue is to continuously track the dynamics of the diseases at a fine resolution. To achieve this, the RS systems should have sufficiently high resolution at all the spatial, spectral, and temporal dimensions. Currently, not a single RS system is able to satisfy these requirements. Besides, bad weather is also a major obstacle to the continuous acquisition of optical images. To tackle this issue, it is important to explore the possibility of synergizing high resolution satellite images with unmanned aerial vehicle (UAV) images to construct a successive time-series RS data. Besides, the fusion between optical RS data and radar data is also worthy of attention.

Considering distributions and epidemics of plant diseases is a transnational process, multinational collaboration is important for both research and application. Presently, the lack of sufficient survey data is a bottleneck in the modeling for monitoring plant diseases and pests. Therefore, it is suggested to mobilize data collection during the cultivation processes. For example, farmers or extension officers should be mobilized to record the occurrence and severity of diseases and pests in their managed fields through a smart phone app. Then, it is critical that the pooled data be easily accessible to support efficient data mining and model training with sophisticated algorithms. Here we expect the setup of corresponding international projects and observation networks that allow experiments, data collection, modeling, and ideas shared at a continental or global scale [59].

4 Notes

1. The appropriate sensor should be chosen based on the wavelengths required for the specific application (e.g., multispectral for broader bands, hyperspectral for fine distinctions). We need to ensure that the spatial resolution fits the scale of your study. High spatial resolution is needed for detecting small-scale features, while lower resolution may be sufficient for larger areas.
2. Depending on the area of study, we need to choose from UAVs, aircraft, or satellite platforms. UAVs provide high-resolution, close-range data, while satellites offer broader coverage.
3. Ensure the sensor is calibrated to adjust for variations in light intensity and sensor drift. This is crucial for ensuring consistency across different datasets and times.
4. Optical sensors are affected by cloud cover and fog, which can block or distort the signal. If cloudy conditions are common, consider using radar or LiDAR, which can penetrate clouds.
5. The time of day and sun angle affect how light is reflected and captured. Collect data at consistent times (typically mid-morning to early afternoon) to minimize shadow and lighting variation. Atmospheric water vapor can affect data quality, especially in the infrared and thermal bands. Correct for atmospheric moisture where possible.
6. Calculating spectral indices in remotes sensing is extremely important and we need to use appropriate vegetation, water, or soil indices (e.g., NDVI, EVI, SAVI, etc.) depending on the application. These indices help highlight specific features like vegetation health or moisture levels.
7. Ground truth data (in situ data collection) is essential to validate remote sensing results. This involves physically inspecting the study area and correlating ground observations with remote sensing data. After analyzing the data, compare the results with ground truth information to assess the accuracy and reliability of the remote sensing interpretation. Use data from multiple sources (optical, radar, LiDAR, etc.) to improve accuracy and provide a fuller picture of the environment.

References

1. Savary S, Ficke A, Aubertot J-N, Hollier C (2012) Crop losses due to diseases and their implications for global food production losses and food security. Food Secur 4:519–537
2. Mahlein A-K (2016) Plant disease detection by imaging sensors–parallels and specific demands for precision agriculture and plant phenotyping. Plant Dis 100(2):241–251
3. Owe M, D'Urso G, Gouweleeuw BT, Jochum AM (2012) Remote sensing for agriculture, ecosystems, and hydrology XIV. In: Neale CMU, Maltese A (eds) Proceedings of SPIE,

vol 8531. https://doi.org/10.1117/12.974722

4. Moran MS, Inoue Y, Barnes EM (1997) Opportunities and limitations for image-based remote sensing in precision crop management. Remote Sens Environ 61(3):319–346
5. Sankaran S et al (2010) A review of advanced techniques for detecting plant diseases. Comput Electron Agric 72(1):1–13
6. Seelan SK et al (2003) Remote sensing applications for precision agriculture: a learning community approach. Remote Sens Environ 88(1–2):157–169
7. West JS et al (2003) The potential of optical canopy measurement for targeted control of field crop diseases. Annu Rev Phytopathol 41(1):593–614
8. Martinelli F et al (2015) Advanced methods of plant disease detection. A review. Agron Sustain Dev 35:1–25
9. Sahoo RN, Ray SS, Manjunath KR (2015) Hyperspectral remote sensing of agriculture. Curr Sci:848–859
10. Xie Y, Sha Z, Yu M (2008) Remote sensing imagery in vegetation mapping: a review. J Plant Ecol 1:9–23. https://doi.org/10.1093/jpe/rtm005
11. Terentev A, Dolzhenko V, Fedotov A, Eremenko D (2022) Current state of hyperspectral remote sensing for early plant disease detection: a review. Sensors 22(3):757
12. Wang YM, Ostendorf B, Gautam D, Habili N, Pagay V (2022) Plant viral disease detection: from molecular diagnosis to optical sensing technology – a multidisciplinary review. Remote Sens 14(7):1542
13. Sankaran S, Mishra A, Ehsani R, Davis C (2010) A review of advanced techniques for detecting plant diseases. Comput Electron Agric 72:1–13
14. Martinelli F, Scalenghe R, Davino S, Panno S, Scuderi G, Ruisi P, Villa P, Stroppiana D, Boschetti M, Goulart RL et al (2015) Advanced methods of plant disease detection: a review. Agron Sustain Dev 35:1–25
15. Oerke E-C, Mahlein A-K, Steiner U (2014) Proximal sensing of plant diseases. In: Gullino ML, Bonants PJM (eds) Detection and diagnostics of plant pathogens, vol 5. Springer, Dordrecht, pp 55–70
16. Jones R (2014) Trends in plant virus epidemiology: opportunities from new or improved technologies. Virus Res 186:3–19
17. Adão T, Hruška J, Pádua L, Bessa J, Peres E, Morais R, Sousa JJ (2017) Hyperspectral imaging: a review on UAV-based sensors, data processing and applications for agriculture and forestry. Remote Sens 9:1110
18. Gautam D, Pagay V (2020) A review of current and potential applications of remote sensing to study the water status of horticultural crops. Agronomy 10:140
19. Aasen H, Honkavaara E, Lucieer A, Zarco-Tejada PJ (2018) Quantitative remote sensing at ultra-high resolution with UAV spectroscopy: a review of sensor technology, measurement procedures, and data correction workflows. Remote Sens 10:1091
20. Hirsch R (2004) Exploring colour photography: a complete guide. Laurence King, London
21. Fu J, Liu J, Zhao R, Chen Z, Qiao Y, Li D (2022) Maize disease detection based on spectral recovery from RGB images. Front Plant Sci 13:1056842
22. Mohanty SP, Hughes DP, Salathé M (2016) Using deep learning for image- based plant disease detection. Front Plant Sci 7:1419
23. Ferentinos KP (2018) Deep learning models for plant disease detection and diagnosis. Comput Electron Agric 145:311–318
24. Wiesner-Hanks T, Wu H, Stewart E, DeChant C, Kaczmar N, Lipson H, Gore MA, Nelson RJ (2019) Millimeter-level plant disease detection from aerial photographs via deep learning and crowdsourced data. Front Plant Sci 10:1550
25. Zhou X-G, Zhang D, Lin F (2021) UAV remote sensing: an innovative tool for detection and management of rice diseases. London, IntechOpen
26. Kang HR (2006) Multispectral imaging. In: Kang HR (ed) Computational color technology. SPIE Press, Bellingham, pp 301–324
27. Ünsalan C, Boyer KL (2011) Multispectral satellite image understanding: from land classification to building and road detection. Springer, London, pp 49–119
28. Curran PJ, Windham WR, Gholz HL (1995) Exploring the relationship between reflectance red edge and chlorophyll concentration in slash pine leaves. Tree Physiol 15:203–206
29. Mulla DJ (2013) Twenty five years of remote sensing in precision agriculture: key advances and remaining knowledge gaps. Biosyst Eng 114:358–371
30. Rouse JW Jr, Harlan JC, Haas RH, Schell JA, Deering DW (1974) Monitoring the vernal advancement and retrogradation (green wave effect) of natural vegetation. NASA/GSFCT Type III Final Report 1974, NASA-CR-144661
31. Davy SH (2013) NDVI from A to Z. In: Pettorelli N (ed) The normalized difference

vegetation index. Oxford University Press, Oxford, pp 30–43

32. Thenkabail PS, Smith RB, De-Pauw E (2002) Evaluation of narrowband and broadband vegetation indices for determining optimal hyperspectral wavebands for agricultural crop characterization. Photogramm Eng Remote Sens 68:607–622
33. Yao X, Zhu Y, Tian Y, Feng W, Cao W (2010) Exploring hyperspectral bands and estimation indices for leaf nitrogen accumulation in wheat. Int J Appl Earth Obs Geoinf 12:89–100
34. Roberts DA, Roth KL, Perroy RL (2011) Hyperspectral vegetation indices. In: Thenkabail PS, Lyon JG, Huete A (eds) Hyperspectral remote sensing of vegetation. CRC Press, Boca Raton
35. Tocqueville AD (2013) Vegetation indices. In: Pettorelli N (ed) The normalized difference vegetation index. Oxford University Press, Oxford, pp 18–29
36. Prabhakar M, Prasad YG, Rao MN (2012) Remote sensing of biotic stress in crop plants and its applications for pest management. In: Venkateswarlu B, Shanker A, Shanker C, Maheswari M (eds) Crop stress and its management: perspectives and strategies. Springer, Dordrecht, pp 517–545
37. Thenkabail PS, Lyon JG, Huete A (2011) Advances in hyperspectral remote sensing of vegetation and agricultural croplands. In: Thenkabail PS, Lyon JG, Huete A (eds) Hyperspectral remote sensing of vegetation. CRC Press, Boca Raton, pp 3–35
38. Griffel L, Delparte D, Edwards J (2018) Using support vector machines classification to differentiate spectral signatures of potato plants infected with potato virus Y. Comput Electron Agric 153:318–324
39. Naidu RA, Perry EM, Pierce FJ, Mekuria T (2009) The potential of spectral reflectance technique for the detection of grapevine leafroll-associated virus-3 in two red-berried wine grape cultivars. Comput Electron Agric 66:38–45
40. Sinha R, Khot LR, Rathnayake AP, Gao Z, Naidu RA (2019) Visible-near infrared spectroradiometry-based detection of grapevine leafroll-associated virus 3 in a red-fruited wine grape cultivar. Comput Electron Agric 162:165–173
41. Afonso AM, Guerra R, Cavaco AM, Pinto P, Andrade A, Duarte A, Power DM, Marques NT (2017) Identification of asymptomatic plants infected with citrus tristeza virus from a time series of leaf spectral characteristics. Comput Electron Agric 141:340–350
42. Grisham MP, Johnson RM, Zimba PV (2010) Detecting sugarcane yellow leaf virus infection in asymptomatic leaves with hyperspectral remote sensing and associated leaf pigment changes. J Virol Methods 167:140–145
43. Qin J (2010) Hyperspectral imaging instruments. In: Sun D-W (ed) Hyperspectral imaging for food quality analysis and control. Academic, San Diego, pp 129–172
44. Gorbe E, Calatayud A (2012) Applications of chlorophyll fluorescence imaging technique in horticultural research: a review. Sci Hortic 138: 24–35
45. Daley PF (1995) Chlorophyll fluorescence analysis and imaging in plant stress and disease. Can J Plant Pathol 17:167–173
46. Osmond CB, Daley PF, Badger MR, Lüttge U (1998) Chlorophyll fluorescence quenching during photosynthetic induction in leaves of Abutilon striatum dicks. Infected with abutilon mosaic virus, observed with a field-portable imaging system. Bot Acta 111:390–397
47. Pineda M, Soukupová J, Matouš K, Nedbal L, Barón M (2008) Conventional and combinatorial chlorophyll fluorescence imaging of tobamovirus-infected plants. Photosynthetica 46:441–451
48. Spoustová P, Synková H, Valcke R, Čeřovská N (2013) Chlorophyll a fluorescence as a tool for a study of the potato virus Y effects on photosynthesis of nontransgenic and transgenic Pssu-ipt tobacco. Photosynthetica 51:191–201
49. Lei R, Jiang H, Hu F, Yan J, Zhu S (2016) Chlorophyll fluorescence lifetime imaging provides new insight into the chlorosis induced by plant virus infection. Plant Cell Rep 36:327–341
50. Chaerle L, Lenk S, Hagenbeek D, Buschmann C, Van Der Straeten D (2007) Multicolor fluorescence imaging for early detection of the hypersensitive reaction to tobacco mosaic virus. J Plant Physiol 164: 253–262
51. Pérez-Bueno ML, Pineda M, Barón M (2019) Phenotyping plant responses to biotic stress by chlorophyll fluorescence imaging. Front Plant Sci 10:1135
52. Fernandez-Jaramillo AA, Duarte-Galvan C, Contreras-Medina LM, Torres-Pacheco I, Romero-Troncoso RDJ, Guevara-Gonzalez RG, Millan-Almaraz JR (2012) Instrumentation in developing chlorophyll fluorescence biosensing: a review. Sensors 12:11853–11869
53. Ni Z, Lu Q, Huo H, Zhang H (2019) Estimation of chlorophyll fluorescence at different scales: a review. Sensors 19:3000

54. Grace J, Nichol C, Disney M, Lewis P, Quaife T, Bowyer P (2007) Can we measure terrestrial photosynthesis from space directly, using spectral reflectance and fluorescence? Glob Chang Biol 13:1484–1497
55. Zarco-Tejada PJ, Jimenez-Berni JA, Suárez L, Sepulcre-Cantó G, Morales F, Miller JR (2009) Imaging chlorophyll fluorescence with an airborne narrow-band multispectral camera for vegetation stress detection. Remote Sens Environ 113:1262–1275
56. MacArthur A, Robinson IC, Rossini M, Davis N, Macdonald K (2014) A dual-field-of-view spectrometer system for reflectance and fluorescence measurements (Piccolo Doppio) and correction of etaloning. In: Proceedings of the fifth international workshop on remote sensing of vegetation fluorescence
57. Chang CY, Zhou R, Kira O, Marri S, Skovira J, Gu L, Sun Y (2020) An unmanned aerial system (UAS) for concurrent measurements of solar- induced chlorophyll fluorescence and hyperspectral reflectance toward improving crop monitoring. Agric For Meteorol 294: 108145
58. Vargas JQ, Bendig J, Mac Arthur A, Burkart A, Julitta T, Maseyk K, Thomas R, Siegmann B, Rossini M, Celesti M et al (2020) Unmanned aerial systems (UAS)-based methods for solar induced chlorophyll fluorescence (SIF) retrieval with non-imaging spectrometers: state of the art. Remote Sens 12:1624
59. Zhang J, Huang Y, Pu R, Gonzalez-Moreno P, Yuan L, Wu K, Huang W (2019) Monitoring plant diseases and pests through remote sensing technology: a review. Comput Electron Agric 165:104943
60. Tartachnyk II, Rademacher I, Kühbauch W (2006) Distinguishing nitrogen deficiency and fungal infection of winter wheat by laser-induced fluorescence. Precis Agr 7:281–293
61. Nicolas H (2004) Using remote sensing to determine of the date of a fungicide application on winter wheat. Crop Prot 23:853–863
62. Khanal S, Fulton J, Shearer S (2017) An overview of current and potential applications of thermal remote sensing in precision agriculture. Comput Electron Agric 139:22–32
63. Singh D, Sao R, Singh KP (2007) A remote sensing assessment of pest infestation on sorghum. Adv Space Res 39:155–163
64. Capodici F, Maltese A, Ciraolo G, D'Urso G, La Loggia G (2017) Power sensitivity analysis of multi-frequency, multi-polarized, multi-temporal SAR data for soil-vegetation system variables characterization. Remote Sens 9:677
65. Butzler TM, Bailey J, Beute MK (1998) Integrated management of sclerotinia blight in peanut: utilizing canopy morphology, mechanical pruning, and fungicide timing. Plant Dis 82:1312–1318

Chapter 36

Electrochemical Methods for Virus Detection

Shambhavi Singh, Sumiah Wani, Nulevino Iralu, Asif Bashir Shikari, and Aflaq Hamid

Abstract

Electrochemical methods have gained significant attention for the rapid, sensitive, and cost-effective detection of plant viruses, which are a major threat to global agriculture. These techniques combine electrochemical transducers with biological recognition elements such as antibodies, aptamers, and nucleic acids to detect plant virus particles or their genetic material with high specificity. Advances in nanotechnology, including the use of materials like graphene, gold nanoparticles, and carbon nanotubes, have significantly enhanced the sensitivity and selectivity of electrochemical biosensors. Commonly used techniques such as amperometry, voltammetry, and electrochemical impedance spectroscopy enable real-time detection, even at very low concentrations of viral biomarkers. These biosensors offer numerous advantages over traditional detection methods, including portability, speed, and the potential for point-of-care applications in the field. This chapter explains recent developments in electrochemical biosensors for plant virus detection, focusing on the integration of nanomaterials, sensor design, and their application in diagnosing economically important plant viruses. It also discusses the challenges in scaling up these technologies for widespread agricultural use and explore future prospects for improving detection efficiency through the development of hybrid systems and advanced sensing platforms. Electrochemical biosensors present a promising tool for enhancing agricultural sustainability and reducing the impact of viral diseases on crop yields. In this chapter, the main focus is upon various electrochemical methods of virus detection using advance tools and techniques.

Key words Electrochemical, Antibodies, Virus, Biosensors

1 Introduction

Plant viruses pose a significant threat to agricultural production, jeopardizing food security and causing substantial economic losses. Rapid and accurate detection methods are crucial for effective disease management [1]. Although traditional methods such as PCR and ELISA are effective, they often involve complex sample preparation, costly reagents, and sophisticated equipment.

Electrochemical biosensors have emerged as promising alternatives due to their high sensitivity, specificity, ease of use, and

Aflaq Hamid et al. (eds.), *Detection of Plant Viruses: Advanced Techniques*, Springer Protocols Handbooks,
https://doi.org/10.1007/978-1-0716-4390-7_36,

potential for field applications [2–5]. These techniques convert biological interactions into measurable electrical signals, enabling label-free detection with rapid response times [6–8].

Electrochemical methods stand out for their high sensitivity, rapid response, affordability, and ease of miniaturization [9]. They rely on measuring electrical signals generated by biochemical reactions between viral targets and specific recognition elements, such as antibodies, nucleic acids, or aptamers. These signals can be quantified and correlated with the concentration of viral particles, facilitating direct and real-time detection [10].

The ability to rapidly detect and identify viral pathogens is essential for effective disease management and prevention. Traditional methods, including polymerase chain reaction (PCR), enzyme-linked immunosorbent assays (ELISA) [27], and viral culture, have been widely used for plant virus detection. However, these techniques often demand sophisticated equipment, lengthy processing times, and highly trained personnel, limiting their feasibility in point-of-care (POC) settings [11–13]. Among the various approaches, electrochemical biosensors have garnered significant attention for virus detection. These devices combine the specificity of biological recognition elements with the sensitivity of electrochemical transducers, providing a powerful tool for identifying viral pathogens. Techniques such as amperometry, voltammetry, and impedance spectroscopy are commonly employed in the development of these biosensors [7, 9, 13–15].

This chapter provides a comprehensive overview of electrochemical methods for virus detection, emphasizing the principles behind these techniques, the types of recognition elements used, and recent advancements in the field [16]. The methodology section outlines the step-by-step process of designing, fabricating, and testing an electrochemical biosensor for viral pathogen detection [7, 9, 10, 17–19].

2 Materials

2.1 Electrodes

1. *Working Electrode*: Various electrode materials, including gold, carbon-based electrodes (graphite, graphene, carbon nanotubes), and platinum, are commonly used. Gold electrodes are preferred for their biocompatibility and ease of surface modification [20].
2. *Counter Electrode*: A platinum wire or glassy carbon electrode serves as the counter electrode to complete the electrochemical cell [20, 21].
3. *Reference Electrode*: The Ag/AgCl electrode is typically used as the reference electrode to maintain a stable potential during the electrochemical reactions [22, 23].

2.2 Surface Modifiers and Nanomaterials

1. *Self-Assembled Monolayers (SAMs)*: SAMs of thiolate compounds such as mercaptohexanoic acid (MHA) or mercaptoundecanoic acid (MUA) are used to immobilize bioreceptors on the electrode surface.
2. *Nanomaterials*: Nanomaterials like gold nanoparticles (AuNPs), silver nanoparticles (AgNPs), and carbon nanotubes (CNTs) enhance sensitivity by providing a larger surface area for bioreceptor immobilization and improving electron transfer kinetics.
3. *Graphene Oxide*: Functionalized graphene oxide offers high surface area and good electrical conductivity, improving the sensor's performance [18, 24, 26].

2.3 Bioreceptors

1. *Antibodies*: Monoclonal and polyclonal antibodies specific to viral coat proteins are used for selective virus detection.
2. *Aptamers*: Short single-stranded DNA or RNA aptamers are synthesized to bind specific viral proteins with high affinity.
3. *DNA Probes*: Short DNA sequences complementary to the viral genome are immobilized on the electrode to detect viral nucleic acids [27].

2.4 Electrolytes and Buffers

1. *Phosphate Buffer Saline (PBS)*: A common buffer used to maintain the physiological pH during the electrochemical measurements.
2. *Supporting Electrolyte*: Potassium chloride (KCl) or sodium chloride (NaCl) is added to the electrolyte to ensure proper ionic strength and conductivity [19, 20].

2.5 Redox Probes

1. *Ferrocyanide (/Ferrocyanide ($Fe(CN)_6^{3-}/Fe(CN)_6^{4-}$)*: These redox-active species are frequently used in cyclic voltammetry (CV) and electrochemical impedance spectroscopy (EIS) measurements.
2. *Methylene Blue (MB)*: An alternative redox probe often employed in DNA hybridization-based biosensors [20]

3 Methods

3.1 Selection of Target Virus and Recognition Elements

3.1.1 Target Virus Selection

Choose a specific virus for detection, considering factors such as its prevalence, impact on public health, and availability of reference materials. For instance, Tobacco Mosaic Virus (TMV) is a prime candidate due to its global significance (*see* **Note 1**).

Virus: Tobacco Mosaic Virus (TMV) was selected as the target virus for detection, owing to its global significance and availability of well-characterized reference materials.

Viral Culture: Tobacco Mosaic Virus (TMV) viral particles were obtained from a certified laboratory, where they were cultured under biosafety level 3 (BSL-3) conditions [14, 19].

3.1.2 Selection of Recognition Elements

Antibodies: Select monoclonal or polyclonal antibodies specific to the viral antigen.

Antibodies: Monoclonal antibodies specific to the Tobacco Mosaic Virus (TMV) spike protein were sourced from a commercial supplier (e.g., Sino Biological, USA). These antibodies were chosen for their high specificity and affinity.

Aptamers: If aptamers are used, select nucleic acid sequences that bind specifically to the viral proteins or nucleic acids.

Aptamers: DNA aptamers specific to the Tobacco Mosaic Virus (TMV) spike protein were synthesized using a SELEX (Systematic Evolution of Ligands by Exponential Enrichment) protocol. The sequences were verified via sequencing before use.

Nucleic Acids: For nucleic acid-based detection, select complementary DNA or RNA sequences to the viral genome.

Nucleic Acids: Single-stranded DNA probes complementary to a conserved region of the Tobacco Mosaic Virus (TMV) genome (e.g., the N gene) were designed [7, 18, 23, 26].

3.1.3 Design and Fabrication of the Electrochemical Biosensor

Electrode Material and Design

Choose a suitable electrode material such as gold, carbon, or platinum, based on its conductivity, stability, and compatibility with the recognition element.

Design the electrode in a manner that maximizes surface area and allows efficient immobilization of the recognition elements. This can include microelectrode arrays or nanostructured surfaces.

Electrode Selection: Gold screen-printed electrodes (SPEs) were chosen for their excellent conductivity, biocompatibility, and ease of functionalization. The electrodes had a working electrode diameter of 4 mm.

Surface Area Optimization: The electrode surface was enhanced with gold nanoparticles (AuNPs) to increase surface area and facilitate the immobilization of the recognition elements. AuNPs were synthesized by the citrate reduction method and characterized using UV-Vis spectroscopy [23, 26, 30].

Electrode Preparation

3.1.4 Electrode Cleaning

The electrode surface (e.g., gold) is cleaned by successive sonication in ethanol, acetone, and ultrapure water for 10 min each to remove organic contaminants.

The surface is then electrochemically cleaned by cycling the electrode potential between −0.2 V and +1.5 V in 0.5 M sulfuric acid until a stable cyclic voltammogram is obtained (*see* **Note 2**).

3.1.5 Modification with Self-Assembled Monolayers (SAMs)

The cleaned electrode is immersed in a 1 mM solution of thiolated SAM (e.g., MHA) in ethanol overnight at room temperature to form a stable monolayer on the surface.

After SAM formation, the electrode is rinsed with ethanol and water to remove excess unbound molecules.

3.1.6 Nanomaterial Decoration

To enhance conductivity and provide a high surface area for bioreceptor immobilization, nanomaterials such as AuNPs or graphene oxide are drop-cast onto the SAM-modified electrode.

The electrode is then dried at room temperature [27, 29, 31–33].

Immobilization of Recognition Elements

3.1.7 Surface Functionalization

Functionalize the electrode surface to enable the stable attachment of recognition elements. Common techniques include self-assembled monolayers (SAMs) for gold electrodes or carboxylation for carbon electrodes.

3.1.8 Immobilization Process

Attach the recognition element to the functionalized surface using covalent bonding, physical adsorption, or affinity interactions. For example, thiolated DNA can be attached to a gold electrode via gold-sulfur bonds.

3.1.9 Surface Functionalization

The gold electrode surface was cleaned with piranha solution (3:1 H2SO4:H2O2) for 1 min to remove organic contaminants, followed by rinsing with ultrapure water and drying under nitrogen. The cleaned electrodes were then incubated in a 1 mM ethanolic solution of 11-mercaptoundecanoic acid (MUA) overnight to form a self-assembled monolayer (SAM) [25, 31–33].

3.2 Immobilization Process

3.2.1 Antibody Immobilization

The MUA-functionalized electrodes were activated using a 1:1 mixture of 0.4 M N-ethyl-N′-(3-dimethylaminopropyl) carbodiimide (EDC) and 0.1 M N-hydroxysuccinimide (NHS) for 30 min. Monoclonal antibodies (100 μg/mL) were then immobilized by incubating the electrodes in the antibody solution for 2 h at room temperature. The electrodes were rinsed with phosphate-buffered saline (PBS) and blocked with 1% bovine serum albumin (BSA) to prevent nonspecific binding.

3.2.2 Aptamer Immobilization

Thiolated DNA aptamers (1 μM) were immobilized on the gold surface via gold-sulfur bonds by incubating the electrodes in the aptamer solution overnight at 4 °C. The electrodes were rinsed with PBS to remove unbound aptamers [23, 27, 31, 34].

3.2.3 Nucleic Acid Probe Immobilization

Thiolated DNA probes (1 μM) were similarly immobilized on the gold electrodes and hybridized with complementary target sequences (100 nM) to confirm successful immobilization [6, 15].

3.2.4 Fabrication of the Biosensor

Assemble the electrochemical cell with the working electrode (where the recognition element is immobilized), a reference electrode (Ag/AgCl), and a counter electrode (platinum wire).

Integrate the sensor into a portable device, if required, for point-of-care applications. This may involve miniaturization and incorporation of wireless communication modules for remote monitoring.

Electrochemical Cell Assembly: The biosensor was assembled in a three-electrode configuration with the gold SPE as the working electrode, Ag/AgCl as the reference electrode, and platinum wire as the counter electrode. The electrodes were connected to a portable potentiostat (e.g., PalmSens4, PalmSens BV, Netherlands) for electrochemical measurements [27, 29, 34].

3.3 Electrochemical Detection Techniques

3.3.1 Amperometric Detection

Measure the current produced by the redox reaction between the viral target and the recognition element at a fixed potential.

The current is directly proportional to the concentration of the viral particles, allowing for quantification.

Setup: The biosensor was operated in amperometric mode using the portable potentiostat. A constant potential of +0.3 V versus Ag/AgCl was applied to the working electrode.

Measurement: The current generated upon binding of the target virus to the recognition element was recorded over a period of 60 s. The current response was proportional to the concentration of Tobacco Mosaic Virus (TMV) particles in the sample [28, 33, 35].

3.3.2 Voltammetric Detection

Apply a varying potential to the electrode and measure the resulting current to generate a voltammogram.

The peak current or potential shift can be correlated with the concentration of the viral target.

Setup: Differential pulse voltammetry (DPV) was employed for enhanced sensitivity. A potential range from −0.2 V to +0.6 V versus Ag/AgCl was applied with a pulse amplitude of 50 mV and a scan rate of 50 mV/s.

Measurement: The resulting voltammogram was recorded, and the peak current was correlated with the concentration of the viral target.

3.3.3 Cyclic Voltammetry (CV)

CV is performed by sweeping the electrode potential between two limits (e.g., −0.2 V to +0.6 V) at a specific scan rate (e.g., 50 mV/s) in the presence of the redox probe (e.g., $Fe(CN)_6^{3-}/Fe(CN)_6^{4-}$).

The changes in current response before and after virus binding are recorded, indicating successful virus detection.

3.3.4 Electrochemical Impedance Spectroscopy (EIS)

EIS measures the changes in the charge transfer resistance (Rct) of the redox probe as the virus binds to the bioreceptor on the electrode surface.

A sinusoidal potential perturbation is applied over a frequency range (e.g., 0.1 Hz to 100 kHz), and the impedance is plotted in a Nyquist plot to quantify virus concentration.

3.3.5 Differential Pulse Voltammetry (DPV)

DPV is employed to detect small changes in current related to the virus concentration. A pulsed potential is applied, and the current response is recorded.

The peak current shift is correlated with the virus load in the sample [38, 41].

3.3.6 Electrochemical Impedance Spectroscopy (EIS)

Setup: EIS was performed using a frequency range of 0.1–100 kHz with a 10 mV AC amplitude superimposed on a DC bias of 0.2 V versus Ag/AgCl.

Measurement: The impedance spectra were analyzed, focusing on the charge transfer resistance (R_ct). Changes in R_ct upon virus binding were used to quantify the viral concentration.

Apply a small AC voltage to the biosensor and measure the resulting impedance.

Analyze changes in the impedance spectrum, particularly the charge transfer resistance (R_ct), which increases with viral binding due to blocking of the electron transfer process [19, 32, 39, 40].

3.4 Optimization of the Biosensor

3.4.1 Sensitivity Enhancement

Nanomaterials: Incorporate nanomaterials such as gold nanoparticles, carbon nanotubes, or graphene to enhance the conductivity and surface area of the electrode, thereby increasing sensitivity.

Signal Amplification: Use enzymatic or catalytic amplification strategies to increase the detectable signal. For example, label antibodies with horseradish peroxidase (HRP) for enhanced amperometric detection [42].

Nanomaterials Integration: Gold nanoparticles (AuNPs) were integrated into the electrode design by drop-casting a 10 μL solution of AuNPs (OD 1.0) onto the electrode surface, followed by drying at room temperature. This step was conducted before the immobilization of recognition elements.

Signal Amplification: To enhance the signal in amperometric detection, secondary antibodies conjugated with horseradish peroxidase (HRP) were employed. The electrodes were incubated with the HRP-conjugated secondary antibody (1:1000 dilution) for 1 h, followed by the addition of the substrate (TMB/H2O2) to generate a measurable current [16, 34, 38, 39].

3.4.2 Selectivity and Stability

Selectivity Testing: The biosensor's selectivity was assessed by exposing it to nontarget viral particles (e.g., influenza A virus) and measuring the electrochemical response. Minimal response was observed, confirming the biosensor's selectivity.

Stability Testing: The biosensor was stored at 4 °C in PBS containing 1% BSA and 0.05% sodium azide. Stability was tested over 30 days by measuring the electrochemical response to a known concentration of Tobacco Mosaic Virus (TMV), with acceptable signal retention (>85%) observed over the testing period.

Optimize the immobilization process to minimize nonspecific binding and enhance selectivity.

Conduct stability tests under various environmental conditions (temperature, pH, etc.) to ensure the long-term reliability of the biosensor [23, 33, 34, 39].

3.5 Testing and Validation

3.5.1 Calibration Curve

Prepare standard solutions of the viral target in a relevant buffer or sample matrix.

Measure the electrochemical response for each concentration and plot a calibration curve to determine the linear range and limit of detection (LOD) of the biosensor (*see* **Note 3**).

Preparation: Serial dilutions of Tobacco Mosaic Virus (TMV) viral particles were prepared in PBS to achieve concentrations ranging from 10^2 to 10^6 particles/mL.

Measurement: Each concentration was measured in triplicate using the biosensor. The current (for amperometry) or peak current (for DPV) was plotted against the viral concentration to generate a calibration curve. The limit of detection (LOD) was calculated using the 3σ rule [6, 9, 11, 15, 28].

3.5.2 Sample Preparation

If working with real biological samples (e.g., blood, saliva), develop protocols for sample preparation, including filtration, dilution, or lysis to ensure compatibility with the biosensor.

Clinical Samples: Nasopharyngeal swabs from Tobacco Mosaic Virus (TMV) positive and negative patients were collected and processed in a biosafety cabinet. Samples were lysed using a commercially available lysis buffer (e.g., TRIzol, Invitrogen) and diluted 1:10 in PBS before testing [22, 13, 35].

3.5.3 Testing with Real Samples

Test the biosensor with real clinical samples to evaluate its performance in detecting the target virus.

3.6 Virus Detection in Plant Samples

3.6.1 Sample Preparation

Leaf samples are collected from suspected infected plants. The samples are ground in PBS or a virus extraction buffer (e.g., 10 mM Tris-HCl, pH 7.5).

The homogenate is filtered or centrifuged to remove debris, and the supernatant is used as the test sample [8, 19, 27].

3.6.2 Sensor Application

The prepared sample is applied to the sensor surface, and the biosensor is incubated for a specific period (e.g., 30 min) to allow virus binding.

The sensor is then rinsed with PBS to remove unbound material before performing electrochemical measurements.

Compare the results with established methods (e.g., PCR) to assess the accuracy, sensitivity, and specificity of the biosensor [42].

Procedure: Processed clinical samples were applied to the biosensor, and the electrochemical response was measured. The results were compared to those obtained by RT-qPCR to assess the biosensor's sensitivity, specificity, and accuracy [9, 20, 30].

3.6.3 Reproducibility and Repeatability

Perform multiple tests with different batches of biosensors to evaluate reproducibility.

Conduct repeated measurements with the same biosensor to assess repeatability.

Reproducibility: The biosensor was fabricated in batches, and the electrochemical response to a standard concentration of Tobacco Mosaic Virus (TMV) was measured across different batches. The coefficient of variation (CV) was calculated to assess reproducibility.

Repeatability: The same biosensor was used for multiple measurements of a single sample, with the CV used to assess repeatability [11, 28, 36].

3.6.4 Data Analysis

Analyze the data using appropriate statistical methods to determine the biosensor's performance characteristics, including sensitivity, specificity, accuracy, precision, and robustness (*see* **Note 4**).

Statistical Methods: Data were analyzed using GraphPad Prism software. Sensitivity, specificity, and accuracy were calculated using standard formulas, and a Bland-Altman plot was generated to compare the biosensor results with RT-qPCR [31, 38, 43].

3.6.5 Validation and Standardization

Validate the biosensor according to international standards (e.g., ISO) for diagnostic devices.

If applicable, conduct a clinical trial to further assess the biosensor's efficacy in a real-world setting.

ISO Standards: The biosensor was validated following ISO 13485 guidelines for in vitro diagnostic devices, including analytical and clinical performance evaluations [36, 44, 45].

3.7 Integration into a Point-of-Care Device

3.7.1 Device Design

Design a user-friendly interface for the biosensor, ensuring it is easy to operate by nonexperts.

Include components such as microfluidic channels for sample handling and a display unit for result interpretation [37].

Interface: A user-friendly interface was developed using a microcontroller (e.g., Arduino) connected to the potentiostat. The interface included a touch screen for sample input and result display.

Sample Handling: Microfluidic channels were integrated into the device to allow automatic sample processing and delivery to the biosensor [11, 15, 26] (*see* **Note 5**).

3.7.2 Power Supply and Signal Processing

Integrate a portable power supply, such as batteries or a solar panel, to ensure the device‘s operability in remote locations.

Develop a signal processing unit that converts the electrochemical signal into a readable format (e.g., digital display or mobile app).

Power Supply: The device was powered by a rechargeable lithium-ion battery (3.7 V, 2000 mAh), providing sufficient power for portable use. A solar panel was incorporated as an alternative power source.

Signal Processing: The electrochemical signal was processed using an onboard analog-to-digital converter (ADC) and displayed as a viral load estimate on the screen. Bluetooth connectivity allowed for data transmission to a smartphone application for remote monitoring [4, 7, 12, 20].

3.7.3 Field Testing

Test the integrated device in real-world conditions to evaluate its practicality, reliability, and usability in point-of-care settings.

Gather feedback from users to refine the device design and improve its functionality.

Real-World Testing: The device was tested [21, 46].

4 Notes

1. *Safety and Handling*: Always follow standard laboratory safety protocols, especially when handling biological samples or reagents. Plant viruses, although not harmful to humans, should be handled with care to prevent contamination of plant tissue cultures. Proper use of personal protective equipment (PPE), such as gloves, lab coats, and safety goggles, is essential.
2. *Preparation of Electrochemical Biosensors*: Ensure all electrodes and biosensor components are thoroughly cleaned before use to avoid signal interference. Electrodes should be pretreated with suitable nanomaterials (e.g., graphene, gold nanoparticles) to enhance conductivity and sensitivity for detecting plant virus biomarkers.
3. *Calibration and Control*: Calibrate the electrochemical sensors using known concentrations of the target virus or synthetic standards. Include both positive and negative controls in all experiments to validate the accuracy and reliability of the biosensor's response.
4. *Data Collection and Interpretation*: Record all measurements in real time, and ensure the electrochemical parameters (e.g., current, voltage) are set appropriately for the specific virus and biosensor. Analyze the data for any significant deviations and validate results through multiple trials.

5. *Storage and Maintenance of Electrochemical Sensors:* Store biosensors in a dry, cool environment to maintain their functionality over time. Electrodes should be stored separately from other reagents to prevent contamination and should be routinely tested for consistency in performance before each experimental run.

References

1. Afsahi S, Lerner MB, Goldstein JM, Lee J, Tang X, Bagarozzi DA et al (2018) Novel graphene-based biosensor for early detection of Zika virus infection. Biosens Bioelectron 100:85–88
2. Ahmadivand A, Gerislioglu B, Ramezani Z, Kaushik A, Manickam P, Ghoreishi SA (2021) Femtomolar-level detection of tobacco mosaic virus (TMV) spike proteins using toroidal plasmonic metasensors. J Mater Chem B 9(33): 7598–7605
3. Arduini F, Cinti S, Scognamiglio V, Moscone D, Palleschi G (2017) How cutting-edge technologies impact the design of electrochemical (bio)sensors for environmental analysis. A review. Anal Chim Acta 959:15–42
4. Chauhan N, Narang J, Pundir CS, Singh M (2017) Fabrication of a highly sensitive and selective aptamer/graphene oxide based electrochemical biosensor for the detection of dengue virus. Biosens Bioelectron 86:372–379
5. Corman VM, Landt O, Kaiser M, Molenkamp R, Meijer A, Chu DK et al (2020) Detection of 2019 novel coronavirus (2019-nCoV) by real-time RT-PCR. Eur Secur 25(3):2000045
6. Dincer C, Bruch R, Kling A, Dittrich PS, Urban GA (2019) Multiplexed point-of-care testing–xPOCT. Trends Biotechnol 37(7): 728–742
7. Farzin L, Sadjadi S, Shamsipur M, Sheibani S (2018) Aptamer-based biosensors for detection of hepatitis C virus. Comb Chem High Throughput Screen 21(2):109–120
8. Kaushik A, Tiwari S, Dev Jayant R, Vashist A, Nair M, Goswami DY, Gupta VK (2016) Electrochemical tobacco mosaic virus (TMV) sensing at point-of-care and artificial intelligence for intelligent Pandmanagement. ACS Appl Bio Mater 3(10):7306–7325
9. Malhotra BD, Ali MA (2017) Nanomaterials for biosensors: fundamentals and applications. William Andrew Publishing
10. Mehmandoust M, Afkhami A, Madrakian T (2020) Gold nanoparticle and nanocomposite based electrochemical sensors for detection of environmental pollutants. Talanta 206:120251
11. Sharma S, Zapatero-Rodríguez J, Estrela P, O'Kennedy R (2015) Point-of-care diagnostics in low-resource settings: present status and future role of microfluidics. Biosens Bioelectron 70:158–173
12. Shen M, Zhou Y, Ye J, Abdullah Al-Maskri AA, Kang Y, Zeng S et al (2020) Recent advances and perspectives of nucleic acid detection for coronavirus. J Pharma Anal 10(2):97–101
13. Torrente-Rodríguez RM, Lukas H, Tu J, Min J, Yang Y, Xu C et al (2020) Tobacco mosaic virus (TMV) rapidPlex: a graphene-based multiplexed telemedicine platform for rapid and low-cost COVID-19 diagnosis and monitoring. Matter 3(6):1981–1998
14. Tripathy S, Singh SG, Singh V, Mahapatra SS (2021) A review on recent developments in electrochemical biosensors for detection of biowarfare agents. Anal Chim Acta 1183: 338867
15. Wang J, Gorski W (2018) Nanomaterial-based biosensors for electrochemical detection of neurotransmitters. Microchim Acta 185:–453
16. Wilson R (2016) The use of gold nanoparticles in biosensors. In: Gold nanoparticles for physics, chemistry, and biology. CRC Press, pp 159–176
17. Xu L, Man Y (2016) Recent advances in the development of electrochemical biosensors for detection of avian influenza virus. Sensors Actuators B Chem 233:255–272
18. Ahmad R et al (2018) Nanomaterials-based electrochemical biosensors for plant virus detection. Biosens Bioelectron
19. Ali MA et al (2019) Graphene-based electrochemical biosensor for virus detection. Electrochim Acta
20. Ahmad R, Tripathy N, Jang N, Hahn YB, Bae H (2017) Highly sensitive biosensor based on field-effect transistor using recombinant

protein for virus detection. Appl Surf Sci 414: 239–243

21. Ahmed RU, Saqib M, Ahmed R, Tahir MA, Masud K (2019) Electrochemical impedance spectroscopy for virus detection: trends and challenges. TrAC Trends Anal Chem 114:47–61
22. Arduini F, Cinti S, Scognamiglio V, Moscone D, Palleschi G (2017) A review of electrochemical biosensors for plant virus detection. TrAC Trends Anal Chem 95:32–45
23. Ballesteros M, Molinero-Fernández Á, Escarpa A, López MA (2018) Electrochemical biosensor for real-time detection of plant virus RNA based on graphene oxide. Biosens Bioelectron 102:343–348
24. Campuzano S, Yanez-Sedeno P, Pingarron JM, Ferapontova EE (2017) Electrochemical DNA biosensors for plant virus detection: from research to application. Sensors Actuators B Chem 253:1051–1062
25. Cantu-Cavada ME, Rodríguez A, Miranda-Contreras L, García J (2021) Advances in the detection of plant pathogens using electrochemical sensors. Microchim Acta 188:326
26. Chen C, Wang J, Li X, Xu Y, Zhao M, Liu Z (2018) Electrochemical biosensor for plant virus detection using nanostructured materials. Electrochim Acta 264:233–239
27. Chou TC, Hsu WL, Li MC (2017) Sensitive detection of plant viruses using enzyme-linked electrochemical biosensors. Biosens Bioelectron 89:587–593
28. Cinti S, Minotti C, Moscone D, Palleschi G, Arduini F (2017) A paper-based electrochemical biosensor for detecting tomato mosaic virus. Sensors Actuators B Chem 252:755–760
29. Corrigan DK, Schulze H, Henihan G, Giraud G, Terry JG, Walton AJ et al (2019) Electrochemical detection of plant viruses using a portable biosensor. Talanta 205: 120097
30. de la Escosura-Muñiz A, Merkoçi A (2016) Nanochannels for DNA sensing by means of electrochemical methods. Anal Chem 88(9): 5061–5066
31. Farzin L, Sadjadi S, Shamsipur M, Sheibani S (2018) Aptamer-based biosensors for detection of plant viruses. Comb Chem High Throughput Screen 21(5):451–461
32. Filippin LI, Bottari NB, Zanella R (2020) Nanomaterial-based electrochemical sensors for plant virus detection. J Electroanal Chem 863:114029
33. Ghosh P, Kumar M, Malhotra BD (2020) Plant virus detection via nanomaterial-based electrochemical biosensors: recent advances. J Mol Recognit 33(12):e2857
34. Hayat A, Marty JL, Sadok S (2017) Plant pathogen detection using electrochemical aptamer-based biosensors. Sensors 17(11):2580
35. Hernández-Barranco A, González-Orive A, Blasco-García A, Escarpa A (2019) Graphene-based electrochemical biosensors for multiplexed detection of plant viruses. Electrochim Acta 319:147–156
36. Hsu HY, Peng PY (2017) Plant virus detection using DNA-functionalized graphene oxide-based electrochemical biosensors. Sensors Actuators B Chem 244:1017–1022
37. Kaushik A, Jayant RD, Vashist A, Nair M, Goswami DY (2018) Electrochemical plant virus sensing at point-of-care and artificial intelligence for precision agriculture. ACS Sustain Chem Eng 6(9):10978–10987
38. Khater M, de la Escosura-Muñiz A, Merkoçi A (2019) Electrochemical biosensor for plant virus detection based on nanomaterials. Anal Chem 91(13):9030–9036
39. Kumar S, Tiwari S, Zain JM, Kaushik A, Nair M, Gupta VK (2016) Nanomaterial-based biosensors for virus detection: recent developments and future perspectives. Biosensors 6(4):49
40. Li B, Zhang P (2019) Biosensing of plant viruses using an electrochemical sensor modified with reduced graphene oxide and gold nanoparticles. Anal Chem 91(10):6835–6841
41. Lim CY, Lee HC, Oh KW, Park TS (2019) Rapid and sensitive detection of plant viruses using a novel electrochemical biosensor. Anal Methods 11(18):2426–2434
42. López-Ruiz N, Sánchez-Montero R, De-Los--Santos-Álvarez N, Lobo-Castañón MJ (2017) Electrochemical aptasensors for plant virus detection based on nanomaterials. Electrochim Acta 247:839–846
43. Medina-Sánchez M, Cadevall M, Merkoçi A (2016) Nanoparticle-based electrochemical biosensors for detection of plant pathogens. Nano Today 11(6):725–737
44. Mehta J, Bhardwaj S, Bhardwaj N, Kim KH, Deep A, Basu S (2018) Recent trends in electrochemical biosensors for detection of plant viruses. Biosens Bioelectron 116:150–164
45. Montes R, Sánchez-Fortún S, Molinero-Fernández Á, Escarpa A (2021) Electrochemical detection of plant viruses using carbon nanodots. Anal Chem 93(1):225–230
46. Pérez N, Torrente-Rodríguez RM, Campuzano S, Pingarrón JM (2016) Nanomaterial-based electrochemical genosensors for the detection of plant pathogens. Biosens Bioelectron 77:1071–1078

Chapter 37

Next-Generation Flow Cytometry Methods for Plant Virus Detection

Insha Mehraj, Fizalah Kawoosa, Sumiah Wani, Nulevino Iralu, Sahar Saleem, and Aflaq Hamid

Abstract

Flow cytometry is a highly effective technique used across various biomedical research fields, including hematology, immunology, and oncology. Its influence extends to plant science, where it has been a critical tool for nearly 40 years, significantly advancing the field. Recent improvements in flow cytometers, along with enhanced calibration and tuning techniques, have made it possible to detect viruses using this method. The rise of flow virometry, a novel use of flow cytometry, allows for the identification, analysis, and characterization of individual viral particles. This innovation holds the potential to uncover new viruses, possibly even discovering viruses akin to the Mimi virus. The chapter outlines the use of flow cytometry and flow virometry for detecting and analyzing plant viruses and viral particles, detailing the materials, methods, and procedural steps required for effective virus detection and characterization.

Key words Flow virometry, Flow cytometry, Plant virus, Viral detection, Centrifugation, Filtration, Fluorescence

1 Introduction

Since the 1970s, cell analysis using FCM and its advanced technique, FACS, have served as essential approaches for examining and isolating individual cells. These techniques have enabled the high-purity separation of rare cellular subpopulations in a single step by detecting and distinguishing cells through light scattering and fluorescence. Fluorescence detection is typically achieved using fluorescent secondary reagents and antibodies targeting cell surface markers or intracellular components in cells with increased membrane permeability [1].

Although FCM has predominantly served in the study of antibiotic susceptibility, cell cycle kinetics, bacterial enumeration, identification of bacterial DNA and fungus spores, and the separation of viable from nonviable bacteria, its application in plant disease

Aflaq Hamid et al. (eds.), *Detection of Plant Viruses: Advanced Techniques*, Springer Protocols Handbooks,
https://doi.org/10.1007/978-1-0716-4390-7_37,

detection remains relatively recent. Combined with fluorescent markers, FCM has served in the swift identification of bacterial pathogens in food, providing precise results in under 30 min at concentrations as low as 10^4 CFU/milliliter. It has also proven effective in identifying bacteria present in soil, like *Bacillus subtilis* in mushroom substrates. Additionally, FCM has been utilized to assess bacterial viability [2].

Since FCM was initially introduced to plant science during the 1980s, its role in biotechnology and breeding has expanded, largely due to the simplicity of sample preparation. FCM uses in plant pathology can be categorized into three main areas: genome size estimation, pathogen identification, and assessment of physiological condition. Genome size estimation involves comparing the fluorescence emitted by DNA labeled with an intercalating dye to a reference with a known genome size. Pathogen identification is achieved through the general staining of nucleic acids or by analyzing scatter patterns and autofluorescence. Although these techniques allow for the identification and quantification of pathogens, they may not differentiate between morphologically similar organisms, requiring the use of specific probes such as antibodies or nucleic acid probes. Assessment of physiological condition is carried out by measuring fluorescence intensity at specific wavelengths, which provides information on cellular metabolic activity. Examples include evaluating membrane integrity with targeted probes, measuring esterase or mitochondrial activity, or detecting GFP (green fluorescent protein) expression [3].

Flow cytometry analysis revealed that leaves from commercial seed varieties "Burpee Hybrid," "Hales Best Jumbo," and "Topmark" were diploid. In contrast, tissue from the proximal regions of germinated cotyledons exhibited a mixoploid nature, showing diploid, tetraploid, and octoploid states. These variations in ploidy levels between cotyledons and leaf tissues are consistent with findings reported for *Cucumis sativus* [4]. Also, FCM has proven to be an effective alternative and complementary technique to chromosome counting for investigating cytogenetic instabilities linked to somaclonal variation in plants regenerated from both long- and short-term embryogenic cell suspension (ECS) cultures, as well as from cultures involving a callus phase or high multiplication rates [5].

Recent advancements in FCM now allow for the detection of microparticles varying in size from 100 nm to 1000 nm. These microparticles could be cellular components, such as exosomes, or external particles like small microbes or viruses. Flow virometry is the application of FCM to identify viral particles [6].

Flow virometry is a quick, flow-based method that doesn't rely on visualization and is less dependent on the operator compared to EFM. It can process between 2000 and 6000 particles per second and provide results within an hour. Using a viral flow cytometer is generally simpler compared to techniques such as TEM, due to the automation of numerous steps. Conventional methods for virus analysis are transmission electron microscopy, ELISA, epifluorescence imaging, western blotting, titration tests, and PCR. These involved bulk analysis and required extensive sample preparation. In contrast, flow virometry enables the direct identification and characterization of individual virus particles.

[7]. Also, viruses from various families (Podoviridae, Baculoviridae, Myoviridae, Herpesviridae, Phycodnaviridae, Retroviridae, Picornaviridae, and Siphoviridae) were analyzed with a standard flow cytometer, marking the first documented instance of directly detecting and differentiating a broad range of viruses using this technique. Additionally, flow virometry is capable of simultaneously identifying multiple plant viruses (tomato mosaic virus (ToMV), potato virus Y (PVY), and cucumber mosaic virus (CMV),) in a single sample [8].

The Luminex MagPlex-TAG bead system, a microsphere-based flow cytometric platform developed by Luminex (Austin, TX, USA), is designed for multiplex nucleic acid detection. This technology has demonstrated its effectiveness in detecting a variety of viruses simultaneously. The Luminex bead-based array has been successfully used to identify multiple types of cassava viruses, begomoviruses, potato viruses, pospi viroids, lily viruses, and grapevine viruses [9]. As flow virometric analytical tools are still new, their applications are developing and expanding gradually.

However, flow virometry has certain limitations, due to the possibility of overestimating the number of particles because of background interference and underestimating counts as a result of coincidence events [10]. Proper dilution is essential to address these challenges. Coincidence events are described by Poisson distribution:

$$P(n) = (rt)ne - (rt)n!$$

Where r is the flow rate (particles per second) and t is the time in the detection volume. A concentration of about 6 × 10^8 particles/mL typically results in a 10 percent chance of two particles being detected simultaneously, with higher concentrations increasing the probability of coincidence. Flow virometry can detect particles down to around 100 nm, which limits its ability to detect smaller virus particles [10].

2 Materials

1. Plant tissue samples (leaves, stems, or roots).
2. Viral transport media (VTM).
3. Blender/Homogenizer.

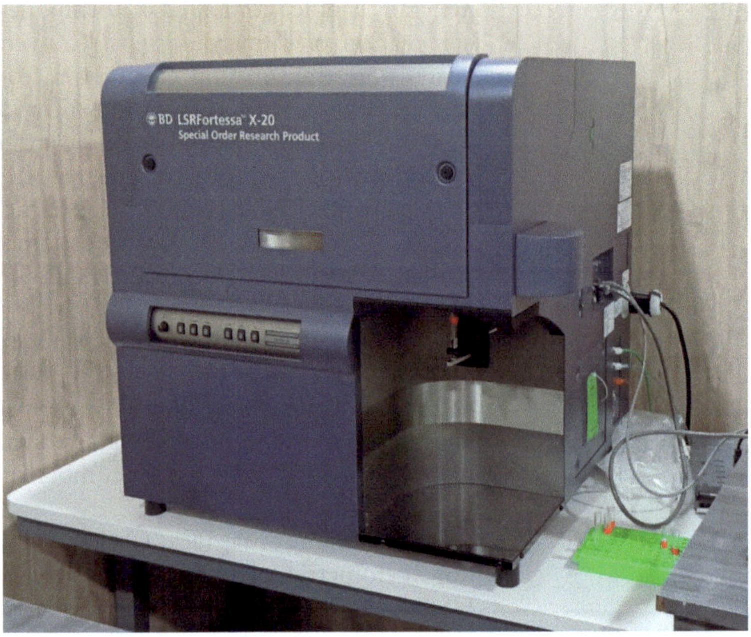

Fig. 1 Flow cytometer (Chapter 37)

4. Ice-cold phosphate-buffered saline (PBS).
5. 0.45-μm cutoff membrane filters.
6. Sodium azide (1% w/v in PBS).
7. Bovine Serum Albumin (BSA, 3% w/v).
8. Primary antibodies specific to plant virus.
9. Fluorochrome-conjugated secondary antibodies.
10. Flow cytometer (Fig. 1).
11. Centrifuge and tubes.
12. 0.1-μm Filtered Sheath Fluid.
13. Vortex.
14. Gloves.

3 Methods

(a) *Collection and Initial Processing:*

(i) *Collect Plant Tissues*: Obtain fresh or preserved plant tissues (e.g., leaves, stems, or roots) suspected of viral infection (*see* **Notes 1** and **17**).

(ii) *Homogenize Tissue*: Grind the collected plant tissues using a blender, homogenizer, or mortar and pestle to create a tissue slurry (*see* **Note 9**).

(iii) *Filter Samples*: Pass the homogenized tissue through a 0.45-μm cutoff membrane to eliminate large particles and minimize aggregation, which helps reduce artifacts and enhance sample clarity (*see* **Note 4**).

(b) *Suspension Preparation:*

(i) *Prepare Suspension*: Resuspend the filtered homogenate in filter-sterilized, ice-cold PBS.

(ii) *Add Sodium Azide (Optional)*: To prevent microbial growth and stabilize viral particles, add sodium azide (1% (w/v) to the PBS solution. This aids in preserving viral surface proteins for better detection (*see* **Note 8**).

(c) *Primary Antibody Incubation:*

(i) *Prepare Primary Antibody Solution*: Dilute 0.1–10 μg of primary antibodies specific to plant viral proteins in 1 ml of PBS suspension. Ensure that the antibodies are appropriate for the plant virus being studied (*see* **Note 3**).

(ii) *Incubation*: Mix well and incubate the tube in the dark for 30–60 min at room temperature. For extended incubation, the sample can be kept at 4 °C in the dark (*see* **Note 10**).

(d) *Washing:*

(i) *Centrifuge and Wash*: Following incubation, centrifuge the sample at 400 g for 5 min at 4 °C. Gently discard the supernatant (*see* **Note 18**).

(ii) *Resuspend Pellet*: Gently resuspend pellet in ice-cold PBS by tapping the tube. Avoid forceful vortexing as it may diminish detection efficiency.

(e) *Secondary Antibody Incubation:*

(i) *Prepare Secondary Antibody Solution*: Dilute Fluorochrome-conjugated secondary antibodies prepared in a 3% (w/v) BSA solution in ice-cold PBS, following the manufacturer's guidelines (*see* **Note 11**).

(ii) *Add to Sample*: In 1 mL of the resuspended virus-antibody mixture from the earlier step, add 0.2–10 μg of secondary antibodies (*see* **Note 7**).

(iii) *Incubation*: Incubate in the dark for at least 30 min at room temperature.

(f) *Final Washing*:

(i) *Centrifuge and Wash*: Wash the sample thrice by centrifugation at 400 g for 5 min using 1 mL of ice-cold PBS with 3% (w/v) BSA and 1% (w/v) sodium azide.

(ii) *Resuspend in PBS*: After the final wash, resuspend the pellet in 100–200 μL of ice-cold PBS for flow cytometry analysis (*see* **Notes 21** and **22**).

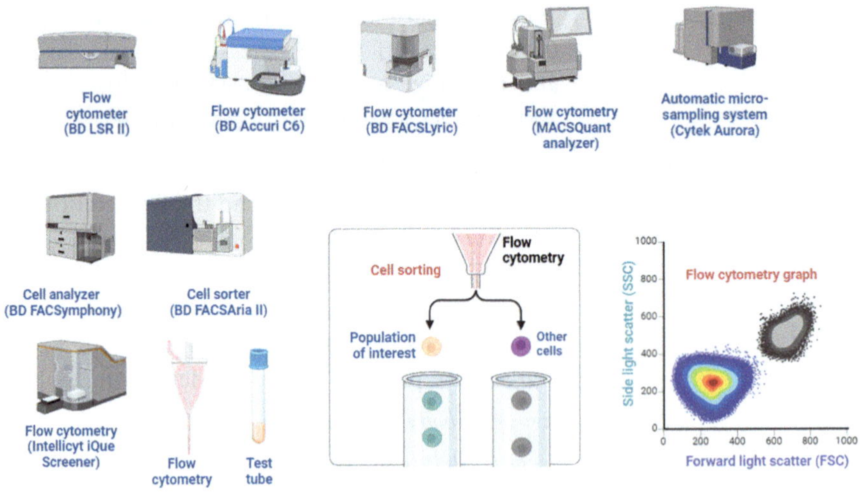

Fig. 2 This image represents flow cytometry results for both control and treated groups, stained with propidium iodide (PI) (Chapter 37). (**a**) Control. (**b**) Treated

(g) *Flow Cytometry Analysis*: (Fig. 2).

  (i) *Prepare Sheath Fluid*: Use a 0.1-μm filter for the sheath fluid to ensure the flow cytometer is well-calibrated for small particles (*see* **Notes 19** and **20**).

  (ii) *Optimize Instrument Settings*: Set the forward scatter (FSC) and side scatter (SSC) thresholds according to the size of the plant virus particles. For example, if detecting smaller plant viruses, adjust FSC and SSC settings to maximize signal-to-noise ratios. Start with default settings and fine-tune based on the specific virus size (*see* **Notes 5** and **6**).

  (iii) *Analyze Samples*: Examine the prepared samples on the flow cytometer promptly to ensure precise readings (*see* **Notes 15** and **16**).

(h) *Controls and Validation:*

  (i) *Blank Control*: Run filtered PBS as a blank to establish background (*see* **Notes 2** and **12**). Fluorescence levels and adjust for any nonspecific signals (*see* **Notes 13** and **14**).

  (ii) *Positive Control*: Use a known positive control sample containing the plant virus with known fluorescence characteristics to validate assay performance (*see* **Note 13**).

  (iii) *Negative Control*: Include healthy plant tissue or untreated samples to confirm that observed signals are specific to the viral particles and not due to background or artifacts (*see* **Note 13**).

These steps will help accurately detect and analyze plant viruses using flow cytometry [11].

4 Notes

1. Use little sample quantities and slower flow rates to enhance virus detection.
2. Combine viruses with calibration beads to provide size references for analysis.
3. To avoid aggregation, use monoclonal antibodies that target single epitopes and a buffer containing EDTA.
4. Preclude viral aggregates by filtering through 0.2 or 0.45 μm size-exclusion filters, minimizing ultracentrifugation steps, and reducing freeze-thaw cycles.
5. Fixing viruses can enhance SSC/FSC and fluorescence signals, but live samples might be ideal for certain viruses and help prevent aggregation.
6. For fluorescent signal detection, determine suitable virus dilutions by plotting various dilutions against mean fluorescence intensity; a horizontal line at $m = 0$ indicates single particle detection [12].
7. For accurate virion detection, distinct viral preparations into two sections and label each with diverse fluorescent dyes, such as maleimide dyes or lipophilic [13].
8. Avoid Contamination: Use sterile equipment and reagents to prevent contamination of samples.
9. Proper Homogenization: Homogenize plant tissues carefully to avoid introducing large particles or debris that can clog the flow cytometer or affect results.
10. Verify Specificity: Ensure that primary and secondary antibodies are specific to the plant virus being studied to avoid cross-reactivity.
11. Handle Fluorochromes Properly: Store fluorochromes according to manufacturer guidelines to prevent degradation. Protect them from light to reduce photobleaching.
12. Daily Calibration: Calibrate the flow cytometer daily using calibration beads to ensure accuracy and consistent performance.
13. Include Controls: Use negative controls (unstained or mock-infected samples) and positive controls (samples with known virus presence) to validate results and correct for background noise or nonspecific signals.
14. Run Replicates: Analyze multiple replicates of each sample to ensure consistency and reliability of the data.

15. Appropriate Software: Utilize flow cytometry analysis software that can handle the specific requirements of virus detection, such as size and fluorescence intensity.
16. Data Interpretation: Carefully interpret data by considering the specific characteristics of virus particles and potential overlaps in fluorescence signals.
17. Personal Protective Equipment (PPE): Wear proper PPE, including lab coats, gloves, and safety glasses, when handling plant tissues, chemicals, and reagents.
18. Chemical Safety: Follow proper protocols for the handling and disposal of chemicals and biological samples to ensure safety and compliance with regulations.
19. Filter Sheath Fluid: Use filtered sheath fluid to prevent contamination and reduce background noise during analysis.
20. Maintain Fluid Levels: Ensure that sheath fluid levels are maintained to prevent air bubbles and ensure smooth flow through the cytometer.
21. Monitor Instrument Performance: Regularly check for issues such as drift in fluorescence signals or changes in scatter characteristics, and address them promptly.
22. Address Artifacts: Be aware of and troubleshoot common artifacts such as clumping, debris, or autofluorescence that can affect results.

References

1. Lippé R (2018) Flow virometry: a powerful tool to functionally characterize viruses. J Virol 92. https://doi.org/10.1128/JVI.01765-17
2. Fang Y, Ramasamy RP (2015) Current and prospective methods for plant disease detection. Biosensors 5:537–561. https://doi.org/10.3390/BIOS5030537
3. D'hondt L, Höfte M, Van Bockstaele E, Leus L (2011) Applications of flow cytometry in plant pathology for genome size determination, detection and physiological status. Mol Plant Pathol 12(8):815–828. https://doi.org/10.1111/J.1364-3703.2011.00711.X
4. Colijn-Hooymans CM, Hakkert JC, Jansen J, Custers JBM (1994) Competence for regeneration of cucumber cotyledons is restricted to specific developmental stages. Plant Cell Tissue Organ Cult 39:211–217. https://doi.org/10.1007/BF00035972
5. Escobedo-GraciaMedrano RM, Maldonado-Borges JI, Burgos-Tan MJ, Valadez-González N, Ku-Cauich JR (2014) Using flow cytometry and cytological analyses to assess the genetic stability of somatic embryo-derived plantlets from embryogenic Musa acuminata Colla (AA) ssp. malaccensis cell suspension cultures. Plant Cell Tissue Organ Cult 116:175–185. https://doi.org/10.1007/S11240-013-0394-Z/METRICS
6. Brown MR, Camézuli S, Davenport RJ, Petelenz-Kurdziel E, Øvreås L, Curtis TP (2015) Flow cytometric quantification of viruses in activated sludge. Water Res 68:414–422
7. Khalil JYB, Langlois T, Andreani J, Sorraing JM, Raoult D, Camoin L, La Scola B (2017) Flow cytometry sorting to separate viable giant viruses from amoeba co-culture supernatants. Front Cell Infect Microbiol 6. https://doi.org/10.3389/FCIMB.2016.00202/FULL
8. D'Hondt L, Höfte M, Van Bockstaele E, Leus L (2011) Applications of flow cytometry in plant pathology for genome size determination, detection and physiological status. Mol Plant Pathol 12:815. https://doi.org/10.1111/J.1364-3703.2011.00711.X

9. Pallás V, Sánchez-Navarro JA, James D (2018) Recent advances on the multiplex molecular detection of plant viruses and viroids. Front Microbiol 9:400508. https://doi.org/10.3389/FMICB.2018.02087/BIBTEX
10. Rossi CA, Kearney BJ, Olschner SP, Williams PL, Robinson CG, Heinrich ML, Zovanyi AM, Ingram MF, Norwood DA, Schoepp RJ (2015) Evaluation of ViroCyt® virus counter for rapid filovirus quantitation. Viruses 7:857–872. https://doi.org/10.3390/V7030857
11. Soni N, Pai P, Kumar GRK, Prasad V, Dasgupta S, Bhadra B (2020) A flow virometry process proposed for detection of SARS-CoV-2 and large-scale screening of COVID-19 cases. Future Virol 15:525–532. https://doi.org/10.2217/FVL-2020-0141
12. Zicari S, Arakelyan A, Fitzgerald W, Zaitseva E, Chernomordik LV, Margolis L, Grivel JC (2016) Evaluation of the maturation of individual dengue virions with flow virometry. Virology 488:20–27. https://doi.org/10.1016/J.VIROL.2015.10.021
13. Arakelyan A, Fitzgerald W, King DF, Rogers P, Cheeseman HM, Grivel JC et al (2017) Flow virometry analysis of envelope glycoprotein conformations on individual HIV virions. Sci Rep 7(1):948. https://doi.org/10.1038/s41598-017-00935-w

Chapter 38

Surface Plasmon Resonance for Virus Detection in Plants

Lubna Javaid, Nulevino Iralu, Sumiah Wani, and Aflaq Hamid

Abstract

Surface plasmon resonance (SPR) is a powerful optical sensing technology that offers real-time, label-free detection of biomolecular interactions, making it a promising tool for viral detection in plants. Plant viruses pose a significant threat to agriculture, leading to reduced crop yields and economic losses worldwide. Conventional methods of viral detection, such as enzyme-linked immunosorbent assays (ELISA) and polymerase chain reaction (PCR), are time-consuming and often require complex sample preparation and specialized equipment. In contrast, SPR provides a rapid and sensitive alternative by directly monitoring the interactions between viral particles and specific biomolecular probes immobilized on a sensor chip. This study explores the application of SPR for detecting various plant viruses, including tobacco mosaic virus (TMV) and potato virus Y (PVY), by using specific antibodies and nucleic acid aptamers. We demonstrate that SPR can achieve high sensitivity and specificity, detecting viral concentrations as low as a few picograms per milliliter. Furthermore, the integration of microfluidic systems with SPR enhances its potential for high-throughput screening of plant samples. The findings highlight the versatility and efficiency of SPR as a diagnostic tool, paving the way for its implementation in agricultural biotechnology to monitor and control viral infections in crops, thereby ensuring food security and sustainable agricultural practices.

Key words Surface plasmon resonance (SPR), Biosensing, Real-time analysis, Label-free detection, Kinetic analysis, Affinity measurements

1 Introduction

Modern virology faces the urgent challenge of rapidly, affordably, and accurately detecting intact viruses and their fragments in various environments, including clinical samples (for human and animal viruses), food products, and infected plant tissues (for plant viruses). Intact viral particles are present during the inactive stage of the viral life cycle. However, their presence in a sample indicates the onset of an infection, and quantifying these particles can help assess the severity of the infection. It's also important to note that similar clinical symptoms can be caused by different viruses. Therefore, developing highly specific and rapid laboratory diagnostic techniques that allow for the direct detection of viruses is crucial.

Aflaq Hamid et al. (eds.), *Detection of Plant Viruses: Advanced Techniques*, Springer Protocols Handbooks,
https://doi.org/10.1007/978-1-0716-4390-7_38,

While many techniques exist to detect virus-specific proteins and nucleic acids, only a few can detect intact viral particles. The traditional method for detecting intact viruses and virus-specific antigens is the enzyme-linked immunosorbent assay (ELISA). However, this technique has certain limitations. One major limitation is that ELISA is based on a thermodynamic process, meaning it provides no information about the kinetic parameters of the reaction. Additionally, ELISA cannot directly identify the source of an antigen, making it challenging to determine whether the response is due to a specific protein, a virus fragment, or the virion itself.

In recent years, nondestructive techniques, such as biosensors, have gained popularity for detecting biospecific interactions, particularly with intact viral particles. For example, the specific detection of bacterial viruses using quartz micro weighing has been demonstrated [1]. Similar approaches can be applied to detect intact human [1] and plant [2] viruses. The bioelectric recognition technique, which relies on the interaction between intact viruses and intact cells, is often used for direct detection of viral particles [3]. Atomic force microscopy is another effective method for studying not only intact viruses but also their developmental stages within a cell [1, 4]. Various optical techniques, such as interferometry [5], can also be employed to detect viral particles in samples due to the higher refractive indices of viral particles compared to biological media.

Optoelectronic transducer techniques offer simplicity and speed, making them ideal for creating convenient and cost-effective instruments for rapid viral infection diagnostics in clinical settings and determining virus content in plants. One of the most efficient techniques widely used for the rapid detection of biospecific interactions is surface plasmon resonance (SPR). This method relies on generating surface polaritons on a thin metal film (such as gold or silver) deposited on the surface of an insulator, as influenced by the surrounding medium [6, 7]. The key advantage of SPR systems over traditional methods for studying immunochemical interactions is their ability to directly investigate intermolecular interactions in real time, as well as the kinetic and thermodynamic parameters of the binding process. This allows for the determination of the mechanism behind these interactions, facilitating the optimization of analytical procedures without the need for labeled reagents [8].

Surface plasmon resonance is extensively used to study virus-specific macromolecules. Most research in this area focuses on individual components rather than entire viral particles [9, 10]. Viral peptides and recombinant or purified viral proteins can serve as convenient models for investigating antibody affinity. However, the structure, conformation, and microenvironment of subunits, both as separate entities and when incorporated into viral particles, can differ significantly. Therefore, studying virus–IgG interactions under conditions that closely mimic in vivo environments is crucial [11]. Viruses are complex mixtures of antigens, with their diversity determined by the number of virus-specific proteins. A single protein often contains multiple antigen determinants, which depend on the molecule's spatial structure and may be located in different regions of the polypeptide chain. Thus, using model viruses with a small number of antigen determinants is essential when developing new methods for detecting intact viruses. The tobacco mosaic virus (TMV) is a convenient model for refining such approaches due to its small size and single envelope protein.

However, surface plasmon resonance has limitations in virology applications. Interpreting results can be challenging for measurements involving intact viruses with lengths (L) of several tens of nanometers. The instrument's resolving power for quantitatively detecting biomolecules (within the linear range, i.e., when $L \ll \lambda$) is determined by the wavelength penetration depth ($\lambda \approx 700$ nm). Therefore, the size of the receptor-analyte complex must not exceed 100 nm for a linear relationship between the angle shift and the total mass of materials on the surface [12]. In its classical form, this technique has been successfully used for detecting specific antibodies with immobilized intact viruses [13], analyzing reversible (low-affinity Langmuir-like process) interactions of viruses with specifically modified surfaces [14], and studying immunoglobulin–virus interactions [1]. Moreover, SPR is well-suited for investigating small plant viruses. However, until now, SPR-based approaches have not been used for the quantitative detection of intact viruses. Therefore, the objective of this study is to develop an SPR approach for the quantitative determination of viruses, using TMV as a model system.

1.1 History of SPR

In 1902, R.W. Wood observed an intriguing phenomenon where dark and light bands appeared in reflected light when polarized light was incident on a mirror with a diffraction grating on its surface. This remarkable physical phenomenon took nearly four decades to be physically interpreted as the excitation of surface plasmons via attenuated total reflection (ATR) [5]. The fundamental principle was based on studies by Zenneck in 1907, who theoretically proved the existence of radio frequency surface electromagnetic (EM) waves at the boundary of a metal (lossy) and dielectric (loss-free) medium, and by Ritchie in 1957, who theoretically demonstrated the existence of surface plasmons at metal surfaces [15]. Subsequent work by Swan and Powell [16] on the excitation of surface plasmons at metallic interfaces with electrons, and later by Stern and Ferrell [14], showed that surface electromagnetic waves at a metallic surface involved the coupling between electromagnetic radiation and surface plasmons.

In its early stages, the unique property of surface plasmon resonance (SPR) being highly sensitive to its boundary conditions was utilized to determine the optical properties of cadmium arachidate thin films deposited on metal surfaces [17]. Nylander demonstrated the utility of SPR in gas sensing. Later, Liedberg and colleagues at the Laboratory of Applied Physics in Linkoping University, Sweden, reported the first biosensing application of SPR. They successfully demonstrated immunosensing of anti-IgG (anti-immunoglobulin G) down to a concentration of 0.2 μg/mL using 60 nm thick silver films deposited on a microscope slide. Since then, a growing number of SPR approaches have been developed in healthcare research, fundamental biological studies, drug design/screening, clinical diagnosis, food and environmental monitoring, defense security, and agricultural insecticide research. Over the years, several generations of SPR biosensing instrumentation have emerged, from prism and grating-coupled devices to fiber-coupled SPR sensors for point-of-care applications.

Current trends in SPR technology focus on improving the sensitivity and accuracy of measurements by applying multilayer graphene or Molybdenum disulfide (MoS2) coatings, enhancing the wear resistance of the sensitive elements, and developing methods for receptor regeneration for multiple uses. SPR imaging with integrated microfluidics lab-on-a-chip (LOC) systems for point-of-care (POC) applications in medical and clinical theranostics appears to be a promising technology for the near future.

Here's a detailed lab manual protocol for using plasmonic sensing for plant virus detection. This protocol provides a step-by-step guide to preparing samples, running the assay, and analyzing the results using Surface Plasmon Resonance.

SPR is used to detect and quantify plant viruses by observing the interaction between viral particles and specific antibodies immobilized on a sensor surface.

1.2 SPR-Based Biosensing

Surface plasmon resonance (SPR) has emerged as a promising optical tool for ultrasensitive detection of bioanalytes, offering an extraordinarily low limit of detection. Compared to other types of electrical and optical sensors, SPR provides the unique advantages of real-time and label-free sensing of biological samples at the molecular level [18, 19]. Over the last few decades, there has been significant progress in the development of SPR sensing devices for detecting various chemical and biological molecules. These devices have applications in a wide range of fields, including medical diagnostics [4], environmental monitoring [20], food safety, and defense security [21].

Researchers are actively exploring the potential of SPR technology to meet the current demand for extremely high sensitivity, with detection capabilities in the nano to picomolar concentration range for very specific targets. This opens up new avenues of research in areas such as drug discovery, food quality, and safety [22, 23]. SPR biosensors are highly sensitive to changes in the refractive index at the metallic sensing substrate, making them valuable for studying the specificity, affinity, and kinetics of biomolecular interactions, as well as for measuring the concentration levels of target analytes in complex matrix samples.

The following section focuses on Kretschmann's configuration in the study of SPR signals. The primary components of this setup include a prism-based optical coupling system, an SPR bio-chip, and an analyte handling system.

1.3 SPR Bio-Chip

Efficient immobilization and capture of bioreceptors on the sensing substrate is a critical step in the development of optical biosensors. The sensor's performance greatly depends on the appropriate surface treatment to ensure effective immobilization of the bioreceptors on the sensor surface. To date, most SPR detection schemes have adopted self-assembled monolayers (SAMs) to conjugate small biomolecules, such as peptides, DNA sequences, and antibodies, to the sensor surface.

Index-matched cover slips are typically used as the substrate for antibody immobilization. After thoroughly cleaning the cover slips, a thin layer of gold, silver, or another metal that supports surface plasmon resonance is sputtered over them, followed by the bio-functionalization of the surface with antibodies.

1.4 Immobilization of Antibody: Functionalization, Activation, and Binding of Receptor

The effective immobilization of antibodies with proper orientation on sensor surfaces is crucial for the performance of immunosensors [21, 24]. During the immobilization process, antibodies must maintain their conformation, and their binding sites must be exposed to antigens to ensure efficient binding. Direct immobilization on sensor surfaces often leads to random orientation of antibodies, making their binding sites inaccessible for specific interactions with target antigens, which decreases binding ability and reduces sensor sensitivity [25, 26].The availability of a highly dense active surface group of immobilized proteins also affects the sensitivity of SPR sensors. Therefore, immobilization protocols must ensure that antibodies are oriented correctly on the sensor surface with exposed active binding sites for the target sample. Various immobilization strategies are reported, which can be classified into three categories: physical, covalent, and bio-affinity immobilization techniques.

Physical adsorption is the simplest method, utilizing ionic bonds or hydrophobic and polar interactions for immobilization. However, this approach results in poor reproducibility due to the random orientation and denaturation of immobilized antibodies [27, 28]. In contrast, covalent and bio-affinity-based immobilization methods provide higher stability and more precisely oriented antibodies.

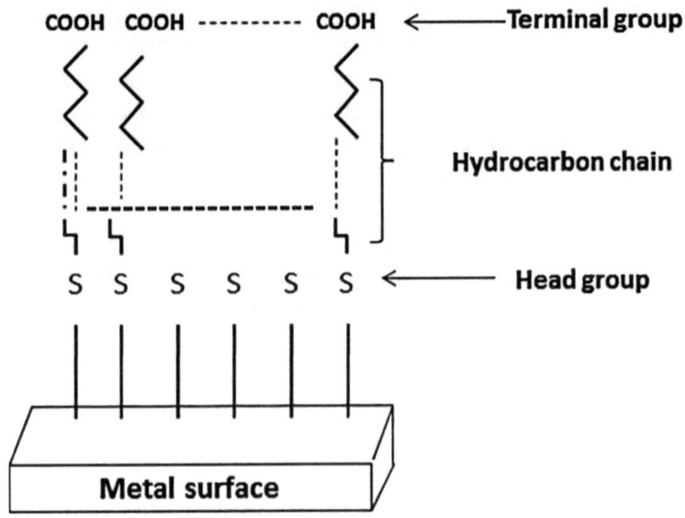

Fig. 1 Representation of SAM of alkanethiols on metal (Chapter 22)

Typically, gold-coated SPR substrates lack sufficient active groups needed for direct antibody attachment [29]. Therefore, inorganic metal surfaces are activated with specific active groups that can efficiently bind to specific moieties (-NH2, -SH, -COOH, etc.) of the protein to be immobilized. Before this activation process, the inorganic surfaces are coated with a protein-resistant self-assembled monolayer (SAM) to maximize the density of active surface groups [26, 30]. SAMs provide stable covalent binding, better control over the orientation of immobilized proteins, and help minimize nonspecific protein adsorption on SPR chips [25, 31]. SAMs are ordered molecular assemblies consisting of an active functional head group at either end of a hydrocarbon chain, with a linear carbon chain promoting self-assembly when attached to a substrate surface. These active head groups strongly bind to the surface, while the other end remains free to bind with antibody moieties.

Monolayers of alkanethiols on gold surfaces are the most extensively studied self-assembled monolayers (SAMs) for protein immobilization [32, 33]. Thiol groups (-SH) have a strong binding affinity to noble metal surfaces, allowing them to form covalent bonds between the sulfur group in the thiol molecules and the metal atoms, as illustrated in Fig. 1. Alkanethiols with acidic end groups offer excellent chemical modification capabilities of the SAM surface, making them promising for sensing applications [34, 35]. SAMs formed from 11-mercaptoundecanoic acid (11-MUA) have been reported to create well-ordered monolayers by various research groups and are used for all the studies reported in this chapter [33].

Table 1
Summary of commonly available functional groups for protein immobilization

Functional group in protein	Amino acid	Substrate requirement	References
Amine (NH2)	Lysine	Carboxylic acid active ester	[37]
Carboxylic (COOH)	Aspartate, glutamate	Amine	[47]
Thiol (-SH)	Cysteine	Maleimide, pyridyl disulfide	[48]
Carbohydrate(-COH)	Sugar residue after oxidation	Silane modified surface	[49]

After surface functionalization for the target antigen, the monolayer surface must be activated to facilitate the covalent attachment of the antibody using moieties such as amine, carboxylic, thiol, and carbohydrate groups available in the protein. Among these functional groups, amine (-NH2) groups in the lysine residues of antibodies are widely used as anchoring points due to their presence on the exterior of the antibody. Table 1 summarizes the common functional groups in antibodies used for covalent attachment and the substrate requirements. The following section presents amine-based chemistry for antibody immobilization, which has been used in all the studies presented here.

This widely used carbodiimide chemistry is applied to form robust amide bonds between the amine groups and the carboxyl groups of antibodies. Water-soluble carbodiimide reagents, such as 1-ethyl-3-(3-dimethylaminopropyl) carbodiimide (EDC), are used to activate the carboxyl groups on the substrate, producing a reactive O-acylisourea intermediate. When N-hydroxy succinimide (NHS) is added, the nucleophile of NHS reacts with the O-acylisourea intermediate, forming an NHS ester that is much more stable than the O-acylisourea intermediate and leads to efficient conjugation to primary amines (James V. Staros, Rick W. Wright, Deborah M. Swingle). The conjugation between aldehyde and amine groups enables the immobilization of antibodies through the formation of a labile Schiff's base.

1.5 Characterization of Sensor Surface

The immobilization protocol for SPR sensors involves several steps, and the sensor surface consists of multiple layers stacked together. The surface composition and structure of these deposited layers, including the thickness of the metal layer, the morphology of the SAM layer for surface functionalization, and the use of NHS-EDC for surface activation, are essential for ensuring the proper orientation of immobilized antibodies. The immobilization of antibodies is typically characterized by spectroscopic techniques such as Fourier-transform infrared spectroscopy (FTIR) and ultraviolet-visible spectroscopy (UV-Vis), as done in our studies and by other research groups. However, various other methods can also be

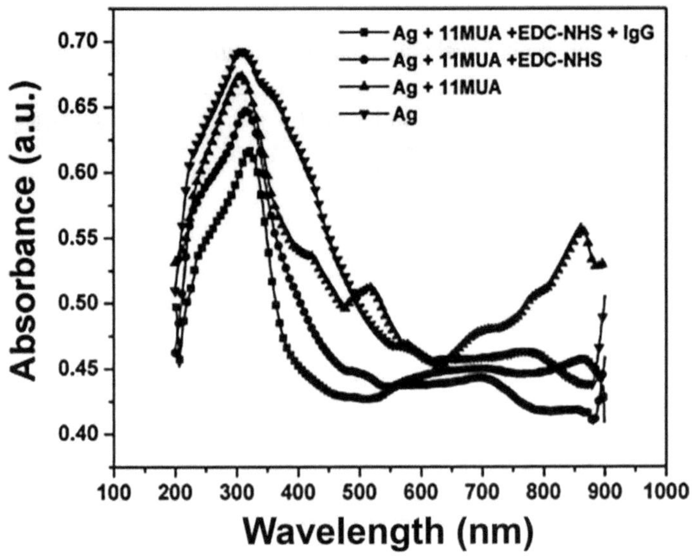

Fig. 2 UV-Vis spectra of the pure silver (Ag) thin film deposited sensor surface before and after antibody (IgG) immobilized; reprinted with permission from Choudhury et al., Creative Commons Attribution 4.0 International License (http:/creativecommons.org/licenses/by/4.0/) (Chapter 22)

employed to assess the state of immobilized antibodies, perform elemental and chemical characterization, and study the surface morphology of the stacked layers. Zhao et al. conducted a detailed study on the surface composition and structure of self-assembled thiol monolayers of alkanethiol molecules using X-ray photoelectron spectroscopy (XPS), FTIR, and energy-dispersive spectroscopy (EDS) [36]. Welch et al. provide a summary of various characterization techniques [26].

Figure 2 shows the optical absorption spectra obtained using UV-visible spectroscopy of the SPR sensing substrate before and after antibody (IgG) immobilization onto a pure silver (Ag) thin-film deposited sensor surface. As observed, the absorbance peak of the pure Ag film is centered at 310 nm, and the subsequent deposition of an additional molecular layer on the silver nanofilm results in a red shift in the SPR dip. The surface functionalization with an 11-MUA layer causes the absorption peak to redshift by approximately 5 nm, attributed to the covalent binding of thiol molecules to the metallic surface. When the antibody is further conjugated, an additional redshift of 15 nm is induced in the absorbance peak. This shift is due to the increase in the local dielectric constant at the metal interface caused by the additional molecular layer.

1.6 SPR Setup

1.6.1 The Optical Assembly

The conventional method for exciting Surface Plasmon Resonance (SPR) uses Kretschmann's configuration, which involves focusing a laser beam through a prism coated with a metal layer. This assembly

Fig. 3 Snap shot of SPR Measurement Set Up (Chapter 23)

is typically placed on a rotating stage. Surface plasmons are excited at the interface between the metal and dielectric layer at a specific incident angle of the light beam. To adjust the incident angle and receive the reflected light, a complex rotation mechanism is required.

Significant advancements in developing portable SPR measurement devices have been made by Prof. Mukherji and his team at the Indian Institute of Technology. Also, several experimental setups, one of which is shown in Fig. 3 [37]. The device includes a BK7 prism mounted with two supports on its sides, allowing easy removal and replacement when necessary. A laser diode (650 nm, 5 mW) is mounted on an "L"-shaped arm, and its movement is controlled by a servomotor, allowing the adjustment of the incident angle of the incoming light within a 45–80° angular range relative to the base of the prism. The incoming light from the laser diode passes through a linear polarizer to generate a p-polarized light beam, which is directed onto the sensor surface at various angles of incidence.

Glycerine, with a refractive index unit (RIU) of 1.42, is used as the index-matching layer to attach microscopic cover slips to the BK7 glass prism, ensuring no air gaps are present between them. A photodiode detector array is employed to collect the light signals reflected from the prism surface. Alternatively, the reflected light from the sensor surface can be focused using a convex lens and collected with a charge-coupled device (CCD) camera connected via a universal serial bus (USB) interface.

1.6.2 Analyte Handling and Sensor Measurement

The device's fluid handling system features a specially designed flow cell, with a volume of approximately 280 mm^3, regulated by a peristaltic pump. This pump is used to pass streams of test samples over the sensor surface. The flow cell can be easily attached to the

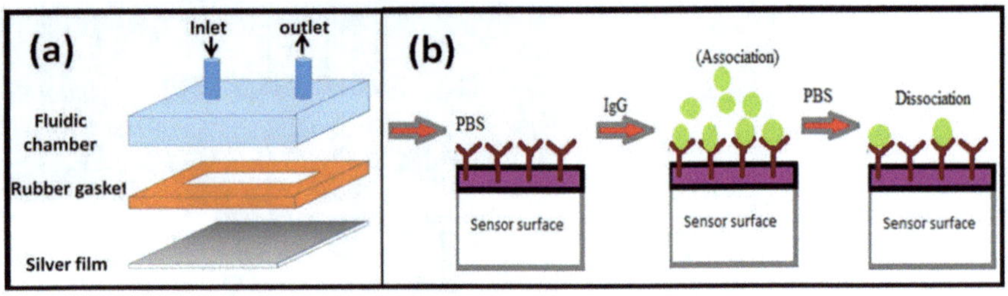

Fig. 4 (**a**) Schematic of fluid handling system and (**b**) sensor measurement procedure (Chapter 23)

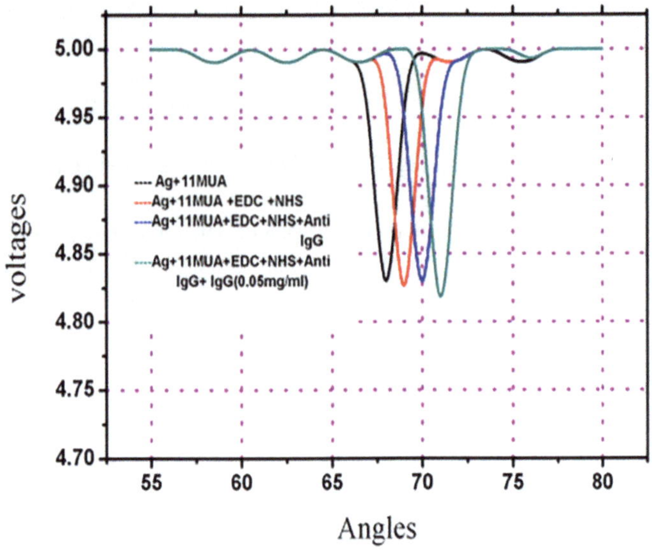

Fig. 5 SPR measurements on the stacking sandwich layers (Chapter 23)

sensor surface using an O-ring, which prevents leakage of the test fluid. A schematic arrangement of the fluid handling system is shown in Fig. 4. The operation of the pumping system is controlled by an Arduino microcontroller.

Before conducting protein conjugation studies, a phosphate buffer solution (PBS) with a pH of 7.4 is flowed over the self-assembled monolayer (SAM to establish baseline signals for subsequent tests. For the measurement steps that follow, the sensor head is exposed to target immunoglobulins (IgGs) in various concentrations, ranging from 0.05 to 0.35 mg/mL, diluted in PBS solution for 10 min. The antigen-antibody binding process is depicted in Fig. 5. In the final step, PBS is injected into the sensing substrate to initiate the dissociation of protein molecules.

1.6.3 Choice of Metal for SPR Sensor

The choice of the metal layer and its thickness significantly impacts the performance characteristics of an SPR sensor. Silver and gold are typically the metals of choice due to their unique properties, such as high efficiency in electron resonance at specific excitation wavelengths, thermal and chemical stability, and the ability to conjugate organic molecules for surface functionalization. Silver films are known to produce a more distinct SPR spectrum than gold. A thin silver film, around 50 nm thick, yields a sharp SPR signal that supports precise determination of the resonance angle [38]. Gold, on the other hand, causes a larger shift in the resonance angle than silver for a fixed change in the refractive index of the sensing layer. Additionally, gold is more stable in ambient conditions and more resistant to oxidation [39].

However, a bimetallic film combining silver and gold can harness the advantages of both materials, leading to improved SPR sensing. Zynio et al. were the first to propose bimetallic films of gold and silver, designing structures where gold was deposited on top of silver to excite SPR. Their findings reported that these bimetallic films exhibited a significant shift in resonance angle due to changes in refractive index, similar to gold films, while maintaining a narrow resonance curve like silver. This combination resulted in high detection accuracy, and the outer gold layer protected the silver from oxidation [40]. Yuan et al. also explored a silver-gold bimetallic configuration. Their results showed that the Full Width at Half Maximum (FWHM) of the bimetallic film was much smaller than that of pure gold films, and the bimetallic film exhibited greater stability than pure silver films [41]. In our recent study, we focused on parameters such as FWHM and sensitivity, examining a set of three bimetallic sensors with varying metal film types and thicknesses [42]. The specifications of these samples are listed in Table 2.

Figure 6 illustrates the SPR responses of sensor chips made from silver, bimetallic silver-gold, and gold when exposed to three test samples: water, acetone, and butanol. The experimental results for the Full Width at Half Maximum (FWHM) obtained from these graphs are detailed in Table 2. The data show that the pure silver sensor chip has the smallest FWHM, indicating the narrowest curve width, which provides the highest resolution for detecting the resonance angle. In contrast, the pure gold-based sensor chip displays the largest curve width among the three configurations. The bimetallic sensor chip, however, demonstrates a notably small

Table 2
experimental results for the Full Width at Half Maximum (FWHM)

Sample ID	Name of the metallic film	Thickness of film
I	Silver	50 nm
II	Gold	50 nm
III	Silver-gold	35(silver) 15(gold)

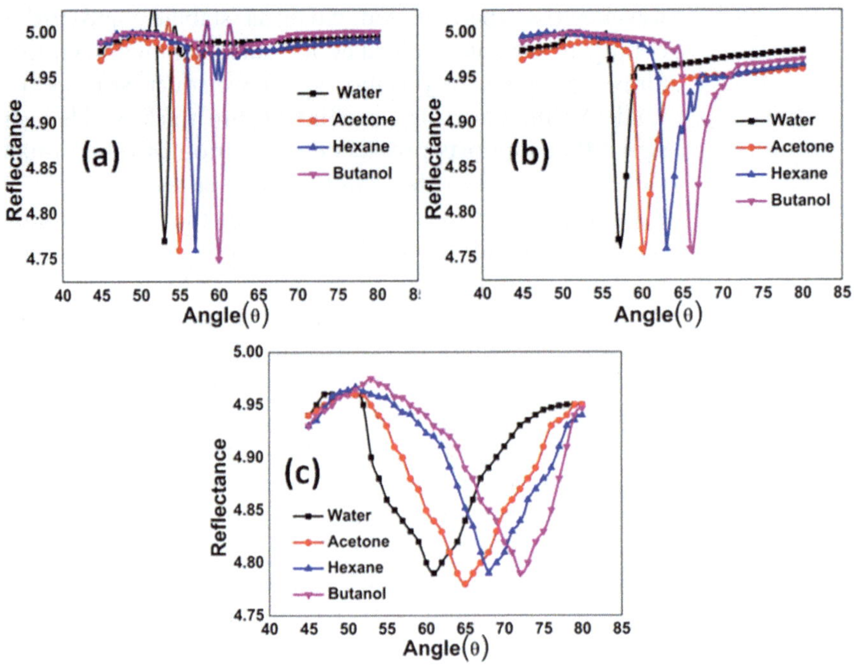

Fig. 6 SPR responses of the silver (**a**), bimetallic (**b**), and gold (**c**) sensor chip toward water, acetone, hexane, and butanol (Chapter 23)

Table 3
Experimental results of FWHM (in degrees) for different configurations

Test samples	Silver (°)	Gold (°)	Silver-gold (°)
Water	1	2	12
Acetone	1	2.5	12
Hexane	1	2	11
Butanol	1	2	11

FWHM—approximately six times smaller than that of the pure gold-based sensor chip—indicating significantly better detection resolution of the resonance angle.

As previously mentioned, the SPR sensitivity is determined by the value of the resonance dip (angular shifts) for a given change in the refractive index of the surrounding sensing media. A larger angular shift corresponds to a more significant signal change, resulting in higher SPR sensitivity. Table 3 presents the experimental measurements of the resonance dip angle shifts (in degrees) due to the changes in the refractive index of the sensing layer, as extracted from the graphs in Fig. 6 (Table 4).

Table 4
Shift of the resonance angle due to change in refractive index of the sensing layer

Refractive index unit (RIU	Shift in resonance angle (in degree) for sensor width		
	Silver	Silver-gold	Gold
0.0261	2°	3.5°	4°

1.7 Advances in SPR Sensors for Biotechnology

In modern biotechnology, particularly in drug discovery, clinical research, and proteomics, the detection of multiple proteins is becoming increasingly important. This requires sophisticated fluidic systems for precise delivery of various fluids to the sensor surface. In this context, Piliarik et al. have reported a highly sensitive SPR sensor that includes an SPR Disperser (SPRCD) and a coupler. The sensor head is integrated with a set of microfluidic channels, enabling multiple measurements [43].

Using this compact system, designed at the Institute of Photonics and Electronics in Prague (Czech Republic), Fernandez et al. developed portable SPR immunosensors with six channels for the simultaneous detection of antibiotics in milk samples [43]. Recently, Liu et al. demonstrated the multi-analyte detection capability of a compact multichannel SPR biosensor, which can even be used on a mobile device as an effective platform for in situ measurements [44].

2 Materials

2.1 Materials

1. Plant virus sample (e.g., Tobacco Mosaic Virus).
2. SPR sensor chips (gold-coated).
3. Specific antibodies against the target plant virus.
4. Phosphate-buffered saline (PBS).
5. Blocking buffer (e.g., BSA solution).
6. Running buffer (e.g., PBS with Tween-20).
7. Regeneration solution (e.g., glycine-HCl, pH 2.5).
8. Pipettes and pipette tips.
9. Micro centrifuge tubes.
10. Distilled water.
11. 96-well plate (optional for sample preparation).

2.2 Equipment

1. Surface Plasmon Resonance (SPR) instrument.
2. Micro centrifuge.
3. Vortex mixer.

4. UV-Vis spectrophotometer (optional for concentration determination).
5. Incubator (optional for incubation steps).

3 Methods

3.1 Immobilization Protocol

The following protocol is a standard process used for the preparation of SPR sensor with immobilized antibody [45, 46].

3.1.1 Substrate Preparation

1. The microscopic cover slips (18×18 mm^2) were cleaned using acetone vapor at 80 °C followed by treatment with methanol at room temperature for 5 min each. Finally, the slides were rinsed in Deionized (DI) water and then dried in a nitrogen environment (*see* **Note 1**).
2. Place the samples onto a hotplate facing up at 180 °C for about 15 min.
3. Sputter deposit ~50 nm gold or silver layer. A thin layer of titanium or chromium (~5 nm) is required to be deposited first before the deposition of Au/Ag layer in order to improve the adhesion capability of Au/Ag with glass.

3.1.2 Functionalization Followed by Activation of Sensor Surface

1. Prepare separate solutions of 1 mM ethanoic solution of 11-MUA, 75 mM EDC, 15 mM NHS, and 50 mM ethanolamine solution in water.
2. Dip the metal-coated substrates in the ethanoic solution 11-MUA in a close container for 24 h at room temperature (*see* **Note 2**).
3. Rinse the MUA-coated chip in ethanol and dry under nitrogen stream.
4. Drop 100 μL of EDC and NHS successively using a micropipette over the sample.
5. Incubate the sample for 1 min in a tightly closed petri dish.

3.1.3 Antibody Immobilization

1. The slide was immersed into anti-mouse IgG2b solutions/IgG in 50 μL phosphate buffer solution (PBS) at pH of 7.4 for 2–4 h to form stable monomolecular layer.
2. The free amine groups of EDC-NHS from the surface functionalization process are blocked through immersing the SPR chip in solutions containing 50 mM ethanolamine (100 μL) for 10–20 min to avoid the unwanted binding of free amine group from EDC-NHS with target molecules.
3. Finally, the chip is washed gently with 0.5 mL PBS and then kept at 4 °C for 12 h (*see* **Note 2**).

3.2 Preparation of SPR Sensor Chip

1. Clean the Sensor Chip:
   - Rinse the gold-coated SPR sensor chip with distilled water.
   - Dry with a gentle stream of nitrogen or air.
2. Immobilization of Antibodies:
   - Dilute the specific antibodies in PBS to the recommended concentration (typically 10–50 μg/mL).
   - Pipette the antibody solution onto the sensor surface and incubate for 1–2 h at room temperature or overnight at 4 °C.
   - Wash the sensor chip with PBS to remove unbound antibodies.
3. Blocking:
   - Apply blocking buffer to the sensor surface to prevent non-specific binding.
   - Incubate for 30 min at room temperature.
   - Rinse with running buffer to remove excess blocking agent.

3.3 Sample Preparation

1. Prepare Plant Virus Sample:
   - If necessary, homogenize plant tissue in PBS using a mortar and pestle or homogenizer.
   - Centrifuge the homogenate at 10,000× *g* for 10 min to remove debris.
   - Collect the supernatant containing the viral particles.
2. Dilute Virus Sample:
   - Prepare a series of dilutions of the plant virus sample in running buffer.
   - Use a spectrophotometer to determine the virus concentration if required.

3.4 Running the SPR Assay

1. Baseline Stabilization:
   - Insert the prepared sensor chip into the SPR instrument.
   - Flow running buffer over the sensor surface to establish a stable baseline.
2. Injection of Plant Virus Sample:
   - Inject the diluted plant virus sample over the sensor chip at a constant flow rate (typically 10–30 μL/min) (*see* **Note 3**).
   - Monitor the SPR angle shift in real time to detect binding events.
3. Regeneration:
   - After sample injection and sufficient interaction time, inject regeneration solution to dissociate bound virus particles.
   - Rinse with running buffer to prepare for the next sample injection (*see* **Note 4**).

3.5 Data Analysis

1. Analyze SPR Data:
   - Use the SPR software to analyze the sensorgram data, which shows the angle shift over time.
   - Determine the association and dissociation rates, as well as the maximum response, to quantify the virus concentration (*see* **Note 5**).
2. Calculate Virus Concentration:
   - Create a standard curve using known concentrations of the virus to determine the concentration in unknown samples.

3.6 Troubleshooting

1. Low Signal:
   - Check antibody immobilization and ensure sufficient coverage on the sensor surface.
   - Verify virus sample concentration and integrity.
2. High Background:
   - Ensure adequate blocking and wash steps to reduce nonspecific binding.
   - Optimize running buffer composition to improve specificity.
3. Irregular Sensorgram:
   - Check for air bubbles in the flow cell.
   - Ensure consistent flow rates and stable baseline before sample injection.

3.7 Safety Precautions

1. Wear appropriate personal protective equipment (PPE), including lab coat, gloves, and safety goggles.
2. Handle all biological samples and reagents according to your institution's safety guidelines.
3. Dispose of waste materials following your laboratory's waste disposal protocols.

4 Notes

1. Choose the appropriate surface chemistry for your specific biomolecules to maximize immobilization efficiency and activity retention.
2. Maintain a consistent temperature throughout experiments to ensure reproducibility and reliable kinetic data.
3. Adjust flow rates to optimize mass transport and avoid artifacts in kinetic measurements.

4. Carefully optimize to ensure complete removal of bound analyte without damaging the immobilized ligand.
5. Ensure proper referencing and baseline subtraction for accurate data analysis.

5 Conclusion

This protocol outlines the steps for using plasmonic sensing via SPR to detect plant viruses, emphasizing the importance of proper sample preparation, antibody immobilization, and data analysis. By following this method, researchers can achieve sensitive and specific detection of plant viruses in various samples.

References

1. Kintzios S, Pistola E, Panagiotopoulos P, Bomsel M, Alexandropoulos N, Bem F et al (2001) Bioelectric recognition assay (BERA). Biosens Bioelectron 16(4–5):325–336
2. Fisch NJ, Malkin VM (2003) Generation of ultrahigh intensity laser pulses. Physics of Plasmas 10(5):2056–2063
3. Eun AJC, Huang L, Chew FT, Li SFY, Wong SM (2002) Detection of two orchid viruses using quartz crystal microbalance (QCM) immunosensors. J Virol Methods 99(1–2): 71–79
4. Homola J (2008) Surface plasmon resonance sensors for detection of chemical and biological species. Chem Rev 108(2):462–493
5. Thome M, Schneider P, Hofmann K, Fickenscher H, Meinl E, Neipel F et al (1997) Viral FLICE-inhibitory proteins (FLIPs) prevent apoptosis induced by death receptors. Nature 386(6624):517–521
6. Snopok BA et al (2003) Quantitative analysis of the molecular ensemble in surface Plasmon resonance. Anal Chem 75:4452–4460
7. Fisch NJ, Malkin VM (2003) Generation of ultrahigh intensity laser pulses. Phys Plsasma 10(5):2056–2063
8. Boltovets PM, Boyko VR, Kostikov IY, Dyachenko NS, Snopok BA, Shirshov YM (2002) Simple method for plant virus detection: effect of antibody immobilization technique. J Virol Methods 105(1):141–146
9. Tanaka Y, Shimoike T, Ishii K, Suzuki R, Suzuki T, Ushijima H et al (2000) Selective binding of hepatitis C virus core protein to synthetic oligonucleotides corresponding to the 5′ untranslated region of the viral genome. Virology 270(1):229–236
10. Snopok BA, Goltsov YG, Kostyukevich EV, Matkovskaja LA, Shirshov YM, Venger EF (2003) Self-assembled multilayer superstructures as immobilization support for bioreceptors. Sens Actuators B Chem 95(1–3): 336–343
11. Cooper MA, Dultsev FN, Minson T, Ostanin VP, Abell C, Klenerman D (2001) Direct and sensitive detection of a human virus by rupture event scanning. Nat Biotechnol 19(9): 833–837
12. Schofield DJ, Dimmock NJ (1996) Determination of affinities of a panel of IgGs and Fabs for whole enveloped (influenza A) virions using surface plasmon resonance. J Virol Methods 62(1):33–42
13. Kuznetsov YG, Datta S, Kothari NH, Greenwood A, Fan H, McPherson A (2002) Atomic force microscopy investigation of fibroblasts infected withWild-type and mutant murine leukemia virus (MuLV). Biophys J 83(6):3665–3674
14. Abad LW, Neumann M, Tobias L, Obenauer-Kutner L, Jacobs S, Cullen C (2002) Development of a biosensor-based method for detection and isotyping of antibody responses to adenoviral-based gene therapy vectors. Anal Biochem 310(1):107–113
15. Kretschmann E, Raether H (1968) Radiative decay of non radiative surface plasmons excited by light. Zeitschrift für Naturforschung A 23(12):2135–2136
16. Ritchie RH (1957) Plasma losses by fast electrons in thin films. Phys Rev 106(5):874
17. Powell CJ, Swan JB (1959) Origin of the characteristic electron energy losses in aluminum. Phys Rev 115(4):869

18. Devine SB, Ferrel RE Jr, Billings GK (1973) The significance of ion exchange to interstitial solutions in clayey sediments. Chem Geol 12(3):219–228
19. Pockrand I, Brillante A, Möbius D (1982) Exciton–surface plasmon coupling: an experimental investigation. J Chem Phys 77(12): 6289–6295
20. Hong DG, Kim TW, Kim KB, Yuk JS, Ha KS (2007) Development of an immunosensor with angular interrogation-based SPR spectroscopy. Meas Sci Technol 18:1367–1371
21. Lijie H, Quentin P, Iban L, Aritz YS, Amaia P, Amaia Z, Daniel M, Rabah B, Sabina S (2017) Label-free femtomolar cancer biomarker detection in human serum using graphene-coated surface plasmon resonance chips. Biosens Bioelectron 89:606–611
22. Mauriz E, Calle A, Manclus JJ, Montoya A, Lechuga LM (2012) Multi-analyte SPR immunoassays for environmental biosensing of pesticides. Anal Bioanal Chem 387:1449–1458
23. Lee JM, Park HK, Jung Y, Kim JK, Jung SO, Chung BH (2007) Direct immobilization of protein G variants with various numbers of cysteine residues on a gold surface. Anal Chem 79:2680–2687
24. Sim HR, Wark AW, Lee HJ (2010) Attomolar detection of protein biomarkers using biofunctionalized gold nanorods with surface plasmon resonance. Analyst 135(10):2528–2532
25. Wang Y, Knoll W, Dostalek J (2012) Bacterial pathogen surface Plasmon resonance biosensor advanced by long range surface Plasmons and magnetic nanoparticle assays. Anal Chem 84(19):8345–8350
26. Qiaoling Y, Wang Q, Li B, Lin Q, Duan Y (2015) Technological development of antibody immobilization for optical immunoassays: Progress and prospects. Crit Rev Anal Chem 45:62–75
27. Kausaite-Minkstimiene A, Ramanaviciene J, Kirlyte A (2010) Ramanavicius, comparative study of random and oriented antibody immobilization techniques on the binding capacity of Immunosensor. Anal Chem 82:6401–6408
28. Welch NG, Scoble JA, Muir BW, Pigram PJ (2017) Orientation and characterization of immobilized antibodies for improved immunoassays (review). Biointerphases 12:02D301
29. Tomizaki K, Usui K, Mihara H (2005) Protein-detecting microarrays: current accomplishments and requirements. Chembiochem 6: 782–799
30. Ma X-X, Liu H-Z, Tang J-J, Guo L, Xie J-W (2011) Comparison of three EnzymeLinked immunosorbent assay methods for quantitative determination of ricin. Chin J Anal Chem 39: 685–689
31. Rusmini F, Zhong Z, Feijen J (2007) Protein immobilization strategies for protein biochips. Biomacromolecules 8:1775–1789
32. Acero Sánchez JL, Fragoso A, Joda H, Suárez G, McNeil CJ, O'Sullivan CK (2016) Site-directed introduction of disulfide groups on antibodies for highly sensitive immunosensors. Anal Bioanal Chem 408:5337–5346
33. Uchida K, Otsuka H, Kaneko M, Kataoka K, Nagasaki Y (2005) A reactive poly (ethylene glycol) layer to achieve specific surface plasmon resonance sensing with a high S/N ratio: the substantial role of a short underbrushed PEG layer in minimizing nonspecific adsorption. Anal Chem 77(4):1075–1080
34. O'Dwyer C, Gay G, Viaris de Lesegno B, Weiner J (2004) The nature of Alkanethiol SelfAssembled monolayer adsorption on sputtered gold substrates. Langmuir 20:8172–8182
35. Stettner J, Winkler A (2010) Characterization of Alkanethiol self-assembled monolayers on gold by thermal desorption spectroscopy. Langmuir 26(12):9659–9665
36. Konwar B (2017) Fabrication of Surface Plasmon Resonance biosensor for snake venom protein detection, Master's Thesis, Department of ECE, Tezpur University
37. Choudhury SN, Konwar B, Kaur S, Doley R, Mondal B (2018) Study on Snake venom protein-antibody interaction by surface Plasmon resonance spectroscopy. Photonic Sens 8(3):193–202
38. Mukherji D, Marques CM, Kremer K (2014) Polymer collapse in miscible good solvents is a generic phenomenon driven by preferential adsorption. Nat Commun 5(1):4882
39. Chakraborty S (2018) Development of surface plasmon resonance based portable refractometer, Master's Thesis, Department of ECE, Tezpur University
40. Nelson SG, Johnston KS, Yee SS (1996) High sensitivity surface plasmon resonance sensor based on phase detection. Sensors Actuators B Chem 35(1–3):187–191
41. Chen X, Jiang K (2010) Effect of aging on optical properties of bimetallic sensor chips. Opt Soc Am 18(2):1105–1112
42. Zynio SA, Samoylov AV, Surovtseva ER, Mirsky VM, Shirshov YM (2002) Bimetallic layers increase sensitivity of affinity sensors based on surface plasmon resonance. Sensors 2(2):62–70
43. Yuan X-C, Ong BH, Tan YG, Zhang DW, Irawan R, Tjin SC (2006) Sensitivity–

stabilityoptimized surface plasmon resonance sensing with double metal layers. J Opt A Pure Appl Opt 8(11):959–963

44. Kashyup R, Chakraborty S, Swarnikar S, Humbert G, Zeng S, Mondal B (2019) FWHM and sensitivity study of bimetallic SPR sensor chip. In: 2019 IEEE international conference on electrical, computer and communication technologies (IEEE ICECCT 2019). SVS College of Engineering, Coimbatore
45. Czanderna AW, King DE, Spaulding D (1991) Metal overlayers on organic functional groups of self-organized molecular assemblies. 1. X-ray photoelectron spectroscopy of interactions of Cu/COOH on 11-mercaptoundecanoic acid. J Vac Sci Technol A 9(5):2607–2613
46. Duevel RV, Corn RM (1992) Amide and ester surface attachment reactions for alkanethiol monolayers at gold surfaces as studied by polarization modulation Fourier transform infrared spectroscopy. Anal Chem 64(4):337–342
47. Fernandez-Lafuente R, Rosell CM, Rodriguez V, Santana C, Soler G, Bastida A, Guisan GM (1993) Enzyme Microb Technol 15:546–550
48. Viitala T, Vikholm I, Peltonen J (2000) Protein immobilization to a partially cross-linked organic monolayer. Langmuir 16(11):4953–4961
49. Qian W, Xu B, Yao D, Lin Y, Wu L, Wang C, Yu F, Lu Z,Wei Y (1999) Site-directed immobilization of immunoglobulin G on 3-aminopropyltriethoxylsilane modified silicon wafer surfaces. Mater Sci Eng C 8:475–480

Chapter 39

In Vivo Synthesis and Purification of Double-Stranded RNA (dsRNA) for RNA Interference (RNAi)

Wani Farhana, Nulevino Iralu, Sumiah Wani, Bikash Mandal, and Aflaq Hamid

Abstract

RNA interference (RNAi) has emerged as an innovative and promising tool for downregulating gene expression, particularly in the context of crop improvement. This advanced technology presents new avenues for developing strategies to knock down and regulate the expression of specific genes, thereby conferring resistance against a variety of plant viruses. This research has uncovered the capacity of double-stranded RNA (dsRNA) molecules to initiate RNAi-mediated gene suppression, offering a means to combat viral diseases in plants and potentially paving the way for the design of RNA vaccines. The threat posed by viruses causing diseases in plants is substantial, impacting global food production and ecological stability. Conventional management systems have proven ineffective in addressing plant viral diseases, necessitating innovative approaches. The concept of RNAi holds significant promise for managing viral diseases in plants. The utilization of the *Escherichia coli* strain HT115 in producing substantial amounts of dsRNAs represents an in vivo bacterial expression system that is not only effective but also more cost-efficient compared to the in vitro transcription method.

Key words RNA interference (RNAi), Double-stranded RNA (dsRNA), *Escherichia coli* strain HT115, RNA vaccines, In vivo bacterial expression system

1 Introduction

Double-stranded RNA plays an important role in regulating the expression of genes in eukaryotes. It induces various types of gene silencing, which are jointly known as RNA interference or RNA silencing [1]. Plants have evolved a mechanism known as post-transcriptional gene silencing (PTGS) to suppress and reduce the expression of certain viral genes [2]. This process includes the sequential degradation of RNA virus and confers plant resistance. The essential characteristic of this process involves the processing of dsRNA into small-interfering RNA (siRNA) of about 21–24 nucleotide long with the aid of DICER-like enzymes [3]. The siRNA binds to the argonaute (AGO) protein before getting introduced

Aflaq Hamid et al. (eds.), *Detection of Plant Viruses: Advanced Techniques*, Springer Protocols Handbooks,
https://doi.org/10.1007/978-1-0716-4390-7_39,

into the RNA-induced silencing complex (RISC), where the viral RNA sequences that coincide with the siRNA sequences are degraded [4]. Moreover, the RNAi mechanism has an amplification loop that generates secondary siRNAs, enhancing the RNAi response. RNA interference-mediated gene silencing has evolved as a potentially effective approach in plants for virus disease resistance [5]. A new advanced molecular approach that uses the topical application of dsRNA to induce antiviral RNA interference is effective against virus infection in plants [6].

While RNAi/gene silencing has demonstrated success in countering viruses, there are notable limitations that impede its widespread application. The effectiveness of RNAi, influenced by factors such as target gene selection, virus titer, sequence similarity, and environmental conditions, is constrained by various interactions [7]. Instances of successful resistance have primarily been observed in controlled environments such as greenhouses, glass houses, or growth chambers, posing challenges for achieving resistance in the presence of mixed infection [8]. A significant limitation lies in the restricted mobility of dsRNAs, hindering their extensive use for crop protection against viruses [9]. Consequently, there is a pressing need for more efficient techniques to facilitate the delivery of dsRNA molecules. Considering these constraints, additional research is essential to pinpoint the factors influencing RNAi-mediated gene silencing, particularly in conferring resistance against plant viruses, and to assess its efficacy under field conditions [10].

2 Materials

1. Trizol reagent (Thermo Fisher Scientific).
2. Chloroform.
3. Isopropanol.
4. cDNA sequence.
5. Revert Aid first-strand cDNA synthesis kit (Thermoscientific).
6. Primers for target cDNA amplification.
7. Agarose.
8. 70% ethanol.
9. 1× TAE (Tris-acetate EDTA) buffer.
10. GoTaq green master mix (Promega).
11. Gel Extraction Kit Promega (Wizard® SV Gel and PCR Clean-Up system).
12. pTZ57R/T (TAVector, PUREGENE).
13. L4440 expression vector (plasmid Addgene 1654A).

14. T4 DNA Ligase (Thermo Fisher Scientific).
15. *Escherichia coli* DH5α competent cells.
16. *E. coli* HT115 competent cells.
17. Ampicillin-100 μg/mL.
18. Tetracyclin-12.5 μg/mL.
19. Plasmid isolation Kit {EndoFree Plasmid Maxi Kit (10)} (Qiagen).
20. Xba I and Hind III restriction enzymes.
21. Enzyme Buffer (10x).
22. IPTG (Isopropyl-β-D-1-thiogalactosid.
23. LB Broth.
24. LB Agar.
25. SDS.
26. 1× PBS.
27. 300 mM sodium acetate.
28. 10 mM Tris–HCl, pH 7.5.
29. 5 mM EDTA.
30. 70% ethanol.
31. RNaseA.
32. Nuclease-free water.
33. Ethidium bromide (EtBr) (Invitrogen).
34. 10× TD loading dye.
35. 1Kb DNA ladder (Thermo Scientific).
36. pTZ57R/T vector (100 ng/μL).
37. 10× T4 DNA Ligase Buffer.
38. T4 DNA Ligase.
39. Insert DNA.
40. 0.1 M $CaCl_2$ (autoclaved).
41. 0.1 M $MgCl_2$ (autoclaved).
42. Sterilized LB Agar plates with proper antibiotics.

3 Methods

3.1 Collect virus-infected samples from field and subject to total RNA isolation using Trizol reagent (Thermo Fisher Scientific) following manufactures protocol (*see* **Note 1**).

3.2 Prepare the complementary DNA using the RevertAid first-strand cDNA synthesis kit (Thermoscientific) according to the manufacturer's instructions.

3.3 Amplify the target sequence by PCR using specific primers using Go GoTaq green master mix (Promega) for adding Poly A tail.

3.4 Purify the PCR-amplified product using Gel Extraction Kit Promega (Wizard® SV Gel and PCR Clean-Up system) following manufacturer's instructions.

3.1 Ligation of PCR-Amplified Target Sequence into pTZ57R/T (TAVector, PUREGENE)

1. Determine the insert DNA to vector DNA ratio. The molar ratio of insert to vector should be between 1:3 for most cloning applications. It is recommended that a series of ligation processes be set up to obtain optimum ligation efficiency, changing the ratio of insert DNA to plasmid DNA.
2. Set up the following reaction in a microcentrifuge tube (*see* **Note 2**).

Table 1: Reaction setup for 20 μL

Component	Volume (μL)
Vector DNA	10.0
10× T4 DNA ligase buffer	2.0
T4 DNA ligase	1.0
Insert DNA	4.0
Nuclease-free water	3.0
Total reaction volume	20.0

Note: The T4 DNA Ligase Buffer should be thawed and resuspended at room temperature. T4 DNA Ligase should be added last

3. Pipette up and down gently to mix the reaction and incubate at 16 °C overnight.
4. Chill on ice and proceed with bacterial transformation.

3.2 Preparation of Competent Cells of Escherichia Coli DH5α

1. Pick a single colony of *E. coli* DH5α or a vial of competent cells from glycerol stock (50–100 mL), inoculate in 5 mL Luria broth (LB) and grown overnight at 37 °C in a BOD shaker incubator (200 rpm) (*see* **Note 3**).
2. Next day, inoculate 250 μL of overnight culture in 200 mL of LB in conical flask and incubate on a rotary shaker at 37 °C (200 rpm) till OD reaches 0.6.
3. Cool the culture on ice and transfer it to 50 ml centrifuge tube.
4. Harvest the cells by centrifugation at 6000× *g* for 10 min at 4 °C. Remove the supernatant carefully and keep the tubes on ice.

5. Resuspend the harvested bacterial cells in 50–70 mL of 0.1 M ice-cold $CaCl_2$ and incubate for 30 min on ice.
6. Centrifuge at 6000× *g* for 10 min at 4 °C to recover the bacterial cells.
7. Resuspend the cells in 50–70 mL of ice-cold 0.1 M $MgCl_2$ and incubated for 30 min on ice.
8. Harvest the cells finally by centrifugation at 6000× *g* for 10 min at 4 °C and suspended in 10 mL of 0.1 M $CaCl_2$ containing 15 percent glycerol.
9. Prepare 200 μL aliquots and store at −80 °C or use directly for transformation.
10. All the steps except the centrifugation should be carried under laminar airflow and the tubes should not be removed from the ice for a longer time.

3.3 Transformation of the Ligated Product into Escherichia Coli DH5α Competent

1. Thaw the frozen competent cells of *E. coli* DH5α stored at −80 °C for 5–10 min on ice.
2. Mix the ligated products with chemically competent *E. coli* DH5α cells and incubated for 30 min on ice.
3. Subject the cells to a 30-s heat shock at 42 °C and immediately move to the ice for at least 5 min (*see* **Note 4**).
4. Add 1 mL of LB (without antibiotic) to the tube and incubate in a shaker incubator for 1 h at 37 °C (200 rpm).
5. After 1 h, harvest the transformation mixture by centrifuging at 5000× *g* for 3 min.
6. Pipette out the top 900 μL of supernatant and mix the remaining 100 μL of supernatant with the small pellet produced by pipetting.
7. Place transformed cells on LB agar plates with the appropriate antibiotics added. Use glass beads to evenly distribute the cells and incubate the plates at 37 °C for 12–16 h.
8. Next day perform colony PCR using specific primers for preliminary screening of transformed cells.

3.4 Isolation of Plasmid from Positive Colonics for Preliminary Screening

1. Pick a single copy of DH5α cell harboring the required vector from a freshly streaked selective plate and transfer into 5 mL LB with antibiotics (5 μL ampicillin) in a tube. Keep the tube in a shaker incubator for 8 h at 37 °C (200 rpm).
2. After incubating for 8 h inoculate 250 μL of grown culture in a fresh 100 mL LB flask. Then grow at 37 °C for 16 h with vigorous shaking (200 rpm) on a shaker incubator.
3. Harvest the bacterial cells by centrifugation at 6000× *g* for 15 min at 4 °C. Discarded the supernatant leaving only the pellet.

4. Add 10 mL of Buffer P1 (RNase A added) to the tube and resuspended the cell pellet completely by vortexing or pipetting up and down until no cell clumps remain.
5. Add 10 mL of Buffer P2 and mix thoroughly by vigorously inverting the sealed tube 4–6 times and incubate at room temperature (25 °C) for 5 min till lysate appears viscous.
6. During the incubation, prepare the QIAfilter Cartridge by screwing the cap onto the outlet nozzle of the QIAfilter Maxi Cartridge. Place QIAfilter Cartridge in a convenient tube.
7. Add 10 mL chilled Buffer P3 to the lysate and mix immediately and thoroughly by vigorously inverting 4–6 times. Mix the suspension until it is colorless. Avoid vortexing.
8. Pour the lysate into the barrel of the QIAfilter Cartridge and incubate at room temperature for 10 min. Don't insert the plunger.
9. Remove the cap from the QIAfilter Cartridge outlet nozzle and insert the plunger gently into the QIAfilter Maxi Cartridge. Filter the cell lysate into a 50 mL tube until all the lysate has passed through the QIAfilter Cartridge. Recover 25 mL of the lysate after filtration.
10. Add 2.5 mL Buffer ER to the filtered lysate and mix by inverting the tube 10 times and incubate on ice for 30 min. After the addition of Buffer ER, the lysate will appear turbid initially but then become clear again during the incubation on ice.
11. Equilibrate QIAGEN-tip 500 by applying 10 mL Buffer QBT. Allow the column to become empty by gravity flow.
12. Add filtered lysate from **step 9** to the QIAGEN-tip and allow to enter the resin by gravity flow.
13. Wash the QIAGEN-tip with 2× 30 mL Buffer QC. Allow the Buffer QC to move through the QIAGEN-tip by gravity flow in a collection tube.
14. Elute the DNA with 15 mL Buffer QN. Collect the eluate in a 30 mL collection tube.
15. Precipitate the DNA by adding 10.5 mL room temperature isopropanol to the eluted DNA. Mix and centrifuge it at 15,000× *g* for 30 min at 4 °C and decant the supernatant.
16. Wash the DNA pellet with 5 mL of 70% ethanol and centrifuge at 15,000× *g* for 10 min at 4 °C. Decant the supernatant carefully without disturbing the pellet.
17. Air-dry the pellet for 5–10 min and redissolve in a suitable volume of endotoxin-free Buffer TE.
18. Determine the DNA concentration by nanodrop and analyze quantitatively on 1% agarose gel.

3.5 Restriction Digestion of Positive pTZ57R/T Clone Using Xba I and HindIII Restriction Sites to Confirm the Clones

Reaction setup (20 μL)	
Digest mixture	**Volume (1×) μL**
Enzyme Buffer (10×)	2
Enzyme XbaI	1
Enzyme hind III	1
Plasmid (pTZ57R/T)	4
Nuclease-free water	12

Timeline reaction setup: 5–10, Incubation: 15–60 min

1. Add appropriate amount of pTZ57R/T plasmid vector to sterile 1.5 mL microcentrifuge tube.
2. Add restriction enzyme Buffer to the tube.
3. Add 1 μL of each Enzyme XbaI and Hind III to the tube.
4. Add nuclease-free water to the tube.

   Important: The enzymes are extremely temperature sensitive. When not in use, they should always be kept on ice. Vortex the enzyme before pipetting to ensure that it is properly mixed.
5. Mix and vortex the reaction mixture one final time.
6. Quickly spin reaction mixture (about 5 s) to collect all liquid in the bottom.
7. Incubate at appropriate temperature (37 °C for our purposes for 1 h (*see* **Note 5**).
8. While the DNA is being digested, prepare a 1% agarose gel.
9. Remove the samples after 60 min of incubation and add 5 μL of tracking dye to sample.
10. Load all 20 μL onto the gel and run at 100 volts for 60 min. At least half of the tracking dye should run down the gel.
11. Examine the gel using a gel documentation system to acquire gel images after being examined on a UV-transilluminator.
12. Purify the gel using Gel Extraction Kit Promega (Wizard® SV Gel and PCR Clean-Up system) following manufacturer's instructions.
    (a) Ligate the eluted insert into L4440 expression vector (having T7 promoter) having same restriction sites in the ratio of 1:3 at 16 °C for overnight using T4 DNA Ligase (Thermo Fisher Scientific) (*see* Subheading 3.1).
    (b) . Next day transform the ligated product into *E. coli* HT115 competent cells (*see* Subheadings 3.2 and 3.3).
    (c) Perform colony PCR next day using specific primers for preliminary screening of transformed cells.

(d) Confirm the clones using restriction digestion (*see* Subheading 3.5).

(e) Use the positive clones for synthesis of dsRNA.

3.6 Culture Preparation and Cell Harvest

1. Pick a single copy of HT115 cell harboring the required insert in the L4440 vector from the master plate and transferred into 5 mL LB with antibiotics (5 μL ampicillin and 5 μL Tetracyclin) in a tube (*see* **Note 6**). Then keep the tube in a shaker incubator overnight at 37 °C (200 rpm) (*see* **Note 7**).
2. The next day, add 250 μL of overnight grown culture to a fresh 50 mL LB flask. Then keep the flask on a shaker incubator for 1 h and check initial (approximately 0.1) (*see* **Note 6**).
3. When the OD_{600} reached 0.4, add 0.5 mM IPTG (Isopropyl-β-D-1-thiogalactoside) to the flask and incubate for 5 h at 37 °C in an incubator [11] (*see* **Note 8**).
4. Distribute the culture equally in oak ridge tubes and harvest the cells by centrifuging for 5 min at 6000× *g* at 4 °C.
5. Discard the supernatant leaving only the cell pellet.

3.6.1 Lysis

1. Resuspended cell pellet is in 2.5 mL of 0.1% SDS (Sodium dodecyl sulfate) that is prepared in 1× PBS (Phosphate-buffered saline) (*see* **Note 9**).
2. Lyse the cells boiling the suspension at 100 °C for 2 min (*see* **Note 10**).
3. After that to remove the ssRNA, give RNaseA treatment in high salt concentration (300 mM sodium acetate, 10 mM Tris–HCl, pH 7.5, and 5 mM EDTA) and then incubate for 30 min at 37 °C (*see* **Note 11**).

3.6.2 Total RNA Isolation Through the Trizol Method

1. Isolate dsRNA from the lysed culture suspension using Trizol reagent according to manufacturer's protocol.

3.6.3 Quality Check of dsRNA

Estimate the quantity of dsRNAs using nanodrop and determine the quality by electrophoresis on 1% agarose gel and store at −20 °C for further.

4 Notes

1. Handle extracted RNA meticulously to prevent contamination and degradation by ensuring cleanliness and sterility of tubes and pipette tips, utilizing certified DNase/RNase-free materials from the isopropanol stage onward, and employing filtered pipette tips for RNA-related transfers and solution preparations.

2. To ensure successful ligation of PCR-amplified target sequences into pTZ57R/T (TAVector, PUREGENE), maintain a sterile environment, optimize reaction conditions, use high-quality PCR products.
3. While preparing competent cells of *E. coli* DH5α, avoid overgrowth of cells during the growth phase prior to transformation. This can be achieved by monitoring cell density and ensuring that cells are harvested at the appropriate growth stage, typically at mid-log phase, to maximize transformation efficiency.
4. Ensure proper heat shock conditions, including precise temperature and duration, to maximize uptake of the plasmid DNA and enhance transformation efficiency.
5. Optimize reaction conditions such as enzyme concentration, Buffer composition, and incubation time to prevent incomplete digestion or nonspecific cleavage, which could lead to inaccurate results.
6. Maintain strict sterility throughout the protocol to prevent contamination during culture preparation and lysis.
7. Ensure accurate antibiotic concentrations to maintain plasmid selection pressure and avoid contamination by non-transformed cells.
8. Monitor OD_{600} accurately for timely IPTG induction, optimizing gene expression without inducing cellular stress.
9. Thoroughly mix the resuspended cell pellet with a properly prepared lysis Buffer for effective cell membrane disruption.
10. Boil the lysed suspension at 100 °C for the specified duration to ensure efficient cell lysis and protein denaturation.
11. Conduct RNaseA treatment in a high salt concentration Buffer to remove residual single-stranded RNA without damaging the double-stranded RNA product.

References

1. Meister G, Tuschl T (2004) Mechanisms of gene silencing by double-stranded RNA. Nature 431:343–349
2. Unver T, Budak H (2009) Virus induced gene silencing, a post-transcriptional gene silencing method. Int J Plant Genomics 2009:1–9
3. Deleris A, Gallego-Bartolome J, Bao J, Kasschau KD, Carrington JC, Voinnet O (2006) Hierarchical action and inhibition of plant Dicer-like proteins in antiviral defense. Science 313:68–71
4. Bologna NG, Voinnet O (2014) The diversity, biogenesis, and activities of endogenous silencing small RNAs in Arabidopsis. Annu Rev Plant Biol 65:473–503
5. Voloudakis AE, Holeva MC, Sarin LP, Bamford DH, Vargas M, Poranen MM, Tenllado F (2015) Efficient double-stranded RNA production methods for utilization in plant virus control. In: Plant virology protocols: new approaches to detect viruses and host responses, pp 255–274
6. Leibman D, Wolf D, Saharan V, Zelcer A, Arazi T, Yoel S, Gal-On A (2011) A high level of transgenic viral small RNA is associated with

broad potyvirus resistance in cucurbits. Mol Plant-Microbe Interact 24:1220–1238

7. Chen W, Zhang X, Fan Y, Li B, Ryabov E, Shi N, Baker B, Wang H, Zhang J, Dong X, Hong Y (2018) A genetic network for systemic RNA silencing in plants. Plant Physiol 176: 2700–2719
8. Dubelman S, Fischer J, Zapata F, Huizinga K, Jiang C, Uffman J, Levine S, Carson D (2014) Environmental fate of double-stranded RNA in agricultural soils. PLoS One 9:e93155
9. Numata K, Ohtani M, Yoshizumi T, Demura T, Kodama Y (2014) Local gene silencing in plants via synthetic dsRNA and carrier peptide. Plant Biotechnol J 12:1027–1034
10. Dalakouras A, Wassenegger M, McMillan JN, Cardoza V, Maegele I, Dadami E, Runne M, Krczal G, Wassenegger M (2016) Induction of silencing in plants by high-pressure spraying of in vitro-synthesized small RNAs. Front Plant Sci 7:1327
11. Gupta D, Singh OW, Basavaraj YB, Roy A, Mukherjee SK, Mandal B (2021) Direct foliar application of dsRNA derived from the full-length gene of NSs of groundnut bud necrosis virus limits virus accumulation and symptom expression. Front Plant Sci 12:734618

INDEX

A

B

C

D

Aflaq Hamid et al. (eds.), *Detection of Plant Viruses: Advanced Techniques*, Springer Protocols Handbooks, https://doi.org/10.1007/978-1-0716-4390-7,

MIX
Papier aus verantwortungsvollen Quellen
Paper from responsible sources
FSC® C105338

If you have any concerns about our products,
you can contact us on
ProductSafety@springernature.com

In case Publisher is established outside the EU,
the EU authorized representative is:
Springer Nature Customer Service Center GmbH
Europaplatz 3, 69115 Heidelberg, Germany

Printed by Libri Plureos GmbH
in Hamburg, Germany